Chemistry: An Introduction

Chemistry: An Introduction

Remi Dalton

STATES
ACADEMIC PRESS
www.statesacademicpress.com

Chemistry: An Introduction
Remi Dalton
ISBN: 978-1-63989-102-3 (Hardback)

Published by States Academic Press,
109 South 5th Street,
Brooklyn, NY 11249, USA

Cataloging-in-Publication Data

Chemistry : an introduction / Remi Dalton.
p. cm.
Includes bibliographical references and index.
ISBN 978-1-63989-102-3
1. Chemistry. 2. Physical sciences. I. Dalton, Remi.
QD31.3 .C44 2022
540--dc23

For more information regarding States Academic Press and its products, please visit the publisher's website www.statesacademicpress.com

Table of Contents

Preface

The scientific study of the character and composition of elements and compounds whether metallic, non-metallic, organic or inorganic is called chemistry. It studies the changes in their structure, behavior and properties whenever they undergo a chemical reaction. The study of energy and entropy is crucial in chemistry as it helps in determining the state of a reaction. Chemical substances involved in these reactions are classified on the basis of their structure, phase and composition. The discipline of chemistry is further divided into various branches such as organic chemistry, inorganic chemistry, nuclear chemistry and analytical chemistry. Different approaches, evaluations and methodologies have been included in this book. It is a valuable compilation of topics, ranging from the basic to the most complex theories and principles in the field of chemistry. For someone with an interest and eye for detail, this book covers the most significant topics in the field of chemistry.

Given below is the chapter wise description of the book:

Chapter 1- The field of enquiry which focuses on the behavior and properties of matter is termed as chemistry. A few of the major areas of study within this field are atoms and molecules, elements and compounds, and substance and mixture. This is an introductory chapter which will introduce briefly these aspects of chemistry.

Chapter 2- There are a number of theories which are studied within chemistry such as atomic theory, molecular orbital theory, transition state theory, Marcus-Hush theory, valence bond theory and VSEPR theory. This chapter has been carefully written to provide an easy understanding of the basic concepts associated with these theories of chemistry.

Chapter 3- The branch of chemistry which deals with the study of organic compounds as well as their structure, reactions and properties is known as organic chemistry. The organic compounds are made up of a carbon atom bonded covalently with other atoms. The diverse reactions which are studied within organic chemistry such as electrophilic substitution reactions, SN1 reaction and SN2 reaction have been thoroughly discussed in this chapter.

Chapter 4- The branch of chemistry which focuses on inorganic and organometallic compounds is termed as inorganic chemistry. It studies their synthesis and behavior. The topics elaborated in this chapter will help in gaining a better perspective about inorganic chemistry, inorganic compounds, inorganic reactions and redox reactions.

Chapter 5- The field of study which deals with macroscopic and particulate phenomena in chemical systems is called physical study. It focuses on their principles and practices as well as the concepts of physics like motion, force, energy, statistical mechanics, etc. All the diverse aspects of physical chemistry such as chemical kinetics, equilibrium chemistry and thermodynamics have been carefully analyzed in this chapter.

At the end, I would like to thank all those who dedicated their time and efforts for the successful completion of this book. I also wish to convey my gratitude towards my friends and family who supported me at every step.

Remi Dalton

1

Basics of Chemistry

The field of enquiry which focuses on the behavior and properties of matter is termed as chemistry. A few of the major areas of study within this field are atoms and molecules, elements and compounds, and substance and mixture. This is an introductory chapter which will introduce briefly these aspects of chemistry.

Chemistry is the science that deals with the properties, composition, and structure of substances (defined as elements and compounds), the transformations they undergo, and the energy that is released or absorbed during these processes. Every substance, whether naturally occurring or artificially produced, consists of one or more of the hundred-odd species of atoms that have been identified as elements. Although these atoms, in turn, are composed of more elementary particles, they are the basic building blocks of chemical substances; there is no quantity of oxygen, mercury, or gold, for example, smaller than an atom of that substance. Chemistry, therefore, is concerned not with the subatomic domain but with the properties of atoms and the laws governing their combinations and how the knowledge of these properties can be used to achieve specific purposes.

The great challenge in chemistry is the development of a coherent explanation of the complex behaviour of materials, why they appear as they do, what gives them their enduring properties, and how interactions among different substances can bring about the formation of new substances and the destruction of old ones. From the earliest attempts to understand the material world in rational terms, chemists have struggled to develop theories of matter that satisfactorily explain both permanence and change. The ordered assembly of indestructible atoms into small and large molecules, or extended networks of intermingled atoms, is generally accepted as the basis of permanence, while the reorganization of atoms or molecules into different arrangements lies behind theories of change. Thus chemistry involves the study of the atomic composition and structural architecture of substances, as well as the varied interactions among substances that can lead to sudden, often violent reactions.

Chemistry also is concerned with the utilization of natural substances and the creation of artificial ones. Cooking, fermentation, glass making, and metallurgy are all chemical processes that date from the beginnings of civilization. Today, vinyl, Teflon, liquid crystals, semiconductors, and superconductors represent the fruits of chemical technology. The 20th century saw dramatic advances in the comprehension of the marvelous and complex chemistry of living organisms, and a molecular interpretation of health and disease holds great promise. Modern chemistry, aided by increasingly sophisticated instruments, studies materials as small as single atoms and as large and complex as DNA (deoxyribonucleic acid), which contains millions of atoms. New substances can even be designed to bear desired characteristics and then synthesized. The rate at which chemical knowledge continues to accumulate is remarkable. Over time more than 8,000,000 different chemical substances, both natural and artificial, have been characterized and produced. The number was less than 500,000 as recently as 1965.

Intimately interconnected with the intellectual challenges of chemistry are those associated with industry. In the mid-19th century the German chemist Justus von Liebig commented that the wealth of a nation could be gauged by the amount of sulfuric acid it produced. This acid, essential to many manufacturing processes, remains today the leading chemical product of industrialized countries. As Liebig recognized, a country that produces large amounts of sulfuric acid is one with a strong chemical industry and a strong economy as a whole. The production, distribution, and utilization of a wide range of chemical products is common to all highly developed nations. In fact, one can say that the "iron age" of civilization is being replaced by a "polymer age," for in some countries the total volume of polymers now produced exceeds that of iron.

The days are long past when one person could hope to have a detailed knowledge of all areas of chemistry. Those pursuing their interests into specific areas of chemistry communicate with others who share the same interests. Over time a group of chemists with specialized research interests become the founding members of an area of specialization. The areas of specialization that emerged early in the history of chemistry, such as organic, inorganic, physical, analytical, and industrial chemistry, along with biochemistry, remain of greatest general interest. There has been, however, much growth in the areas of polymer, environmental, and medicinal chemistry during the 20th century. Moreover, new specialities continue to appear, as, for example, pesticide, forensic, and computer chemistry.

Analytical Chemistry

Most of the materials that occur on Earth, such as wood, coal, minerals, or air, are mixtures of many different and distinct chemical substances. Each pure chemical substance (e.g., oxygen, iron, or water) has a characteristic set of properties that gives it its chemical identity. Iron, for example, is a common silver-white metal that melts at 1,535 °C, is very malleable, and readily combines with oxygen to form the common substances hematite and magnetite. The detection of iron in a mixture of metals, or in a compound such as magnetite, is a branch of analytical chemistry called qualitative analysis. Measurement of the actual amount of a certain substance in a compound or mixture is termed quantitative analysis. Quantitative analytic measurement has determined, for instance, that iron makes up 72.3 percent, by mass, of magnetite, the mineral commonly seen as black sand along beaches and stream banks. Over the years, chemists have discovered chemical reactions that indicate the presence of such elemental substances by the production of easily visible and identifiable products. Iron can be detected by chemical means if it is present in a sample to an amount of 1 part per million or greater. Some very simple qualitative tests reveal the presence of specific chemical elements in even smaller amounts. The yellow colour imparted to a flame by sodium is visible if the sample being ignited has as little as one-billionth of a gram of sodium. Such analytic tests have allowed chemists to identify the types and amounts of impurities in various substances and to determine the properties of very pure materials. Substances used in common laboratory experiments generally have impurity levels of less than 0.1 percent. For special applications, one can purchase chemicals that have impurities totaling less than 0.001 percent. The identification of pure substances and the analysis of chemical mixtures enable all other chemical disciplines to flourish.

The importance of analytical chemistry has never been greater than it is today. The demand in modern societies for a variety of safe foods, affordable consumer goods, abundant energy, and labour-saving technologies places a great burden on the environment. All chemical manufacturing produces waste products in addition to the desired substances, and waste disposal has not always been carried out carefully. Disruption of the environment has occurred since the dawn of civilization, and pollution problems have increased with the growth of global population. The techniques of analytical chemistry are relied on heavily to maintain a benign environment. The undesirable substances in water, air, soil, and food must be identified, their point of origin fixed, and safe, economical methods for their removal or neutralization developed. Once the amount of a pollutant deemed to be hazardous has been assessed, it becomes important to detect harmful substances at concentrations well below the danger level. Analytical chemists seek to develop increasingly accurate and sensitive techniques and instruments.

Sophisticated analytic instruments, often coupled with computers, have improved the accuracy with which chemists can identify substances and have lowered detection limits. An analytic technique in general use is gas chromatography, which separates the different components of a gaseous mixture by passing the mixture through a long, narrow column of absorbent but porous material. The different gases interact differently with this absorbent material and pass through the column at different rates. As the separate gases flow out of the column, they can be passed into another analytic instrument called a mass spectrometer, which separates substances according to the mass of their constituent ions. A combined gas chromatograph–mass spectrometer can rapidly identify the individual components of a chemical mixture whose concentrations may be no greater than a few parts per billion. Similar or even greater sensitivities can be obtained under favourable conditions using techniques such as atomic absorption, polarography, and neutron activation. The rate of instrumental innovation is such that analytic instruments often become obsolete within 10 years of their introduction. Newer instruments are more accurate and faster and are employed widely in the areas of environmental and medicinal chemistry.

Inorganic Chemistry

Modern chemistry, which dates more or less from the acceptance of the law of conservation of mass in the late 18th century, focused initially on those substances that were not associated with living organisms. Study of such substances, which normally have little or no carbon, constitutes the discipline of inorganic chemistry. Early work sought to identify the simple substances namely, the elements that are the constituents of all more complex substances. Some elements, such as gold and carbon, have been known since antiquity, and many others were discovered and studied throughout the 19th and early 20th centuries. Today, more than 100 are known. The study of such simple inorganic compounds as sodium chloride (common salt) has led to some of the fundamental concepts of modern chemistry, the law of definite proportions providing one notable example. This law states that for most pure chemical substances the constituent elements are always present in fixed proportions by mass (e.g., every 100 grams of salt contains 39.3 grams of sodium and 60.7 grams of chlorine). The crystalline form of salt, known as halite, consists of intermingled sodium and chlorine atoms, one sodium atom for each one of chlorine. Such a compound, formed solely by the combination of two elements, is known as a binary compound. Binary compounds are very common in inorganic chemistry, and they exhibit little structural variety. For this reason, the number of inorganic compounds is limited in spite of the large number of elements that may react with each other. If three or more elements are combined in a substance, the structural possibilities become greater.

After a period of quiescence in the early part of the 20th century, inorganic chemistry has again become an exciting area of research. Compounds of boron and hydrogen, known as boranes, have unique structural features that forced a change in thinking about the architecture of inorganic molecules. Some inorganic substances have structural features long believed to occur only in carbon compounds, and a few inorganic polymers have even been produced. Ceramics are materials composed of inorganic elements combined with oxygen. For centuries ceramic objects have been made by strongly heating a vessel formed from a paste of powdered minerals. Although ceramics are quite hard and stable at very high temperatures, they are usually brittle. Currently, new ceramics strong enough to be used as turbine blades in jet engines are being manufactured. There is hope that ceramics will one day replace steel in components of internal-combustion engines. In 1987 a ceramic containing yttrium, barium, copper, and oxygen, with the approximate formula $YBa_2Cu_3O_7$, was found to be a superconductor at a temperature of about 100 K. A superconductor offers no resistance to the passage of an electrical current, and this new type of ceramic could very well find wide use in electrical and magnetic applications. A superconducting ceramic is so simple to make that it can be prepared in a high school laboratory. Its discovery illustrates the unpredictability of chemistry, for fundamental discoveries can still be made with simple equipment and inexpensive materials.

Many of the most interesting developments in inorganic chemistry bridge the gap with other disciplines. Organometallic chemistry investigates compounds that contain inorganic elements combined with carbon-rich units. Many organometallic compounds play an important role in industrial chemistry as catalysts, which are substances that are able to accelerate the rate of a reaction even when present in only very small amounts. Some success has been achieved in the use of such catalysts for converting natural gas to related but more useful chemical substances. Chemists also have created large inorganic molecules that contain a core of metal atoms, such as platinum, surrounded by a shell of different chemical units. Some of these compounds, referred to as metal clusters, have characteristics of metals, while others react in ways similar to biologic systems. Trace amounts of metals in biologic systems are essential for processes such as respiration, nerve function, and cell metabolism. Processes of this kind form the object of study of bioinorganic chemistry. Although organic molecules were once thought to be the distinguishing chemical feature of living creatures, it is now known that inorganic chemistry plays a vital role as well.

Organic Chemistry

Organic compounds are based on the chemistry of carbon. Carbon is unique in the variety and extent of structures that can result from the three-dimensional connections of its atoms. The process of photosynthesis converts carbon dioxide and water to oxygen and compounds known as carbohydrates. Both cellulose the substance that gives structural rigidity to plants, and starch, the energy storage product of plants, are polymeric carbohydrates. Simple carbohydrates produced by photosynthesis form the raw material for the myriad organic compounds found in the plant and animal kingdoms. When combined with variable amounts of hydrogen, oxygen, nitrogen, sulfur, phosphorus, and other elements, the structural possibilities of carbon compounds become limitless, and their number far exceeds the total of all nonorganic compounds. A major focus of organic chemistry is the isolation, purification, and structural study of these naturally occurring substances. Many natural products are simple molecules. Examples include formic acid (HCO_2H) in ants, ethyl alcohol (C_2H_5OH) in fermenting fruit, and oxalic acid ($C_2H_2O_4$) in rhubarb leaves. Other natural products, such as penicillin, vitamin B_{12}, proteins, and nucleic acids, are exceedingly complex. The isolation of pure natural products from their host organism is made difficult by the low concentrations in which they may be present. Once they are isolated in pure form, however, modern instrumental techniques can reveal structural details for amounts weighing as little as one-millionth of a gram. The correlation of the physical and chemical properties of compounds with their structural features is the domain of physical organic chemistry. Once the properties endowed upon a substance by specific structural units termed functional groups are known, it becomes possible to design novel molecules that may exhibit desired properties. The preparation, under controlled laboratory conditions, of specific compounds is known as synthetic chemistry. Some products are easier to synthesize than to collect and purify from their natural sources. Tons of vitamin C, for example, are synthesized annually. Many synthetic substances have novel properties that make them especially useful. Plastics are a prime example, as are many drugs and agricultural chemicals. A continuing challenge for synthetic chemists is the structural complexity of most organic substances. To synthesize a desired substance, the atoms must be pieced together in the correct order and with the proper three-dimensional relationships. Just as a given pile of lumber and bricks can be assembled in many ways to build houses of several different designs, so too can a fixed number of atoms be connected together in various ways to give different molecules. Only one structural arrangement out of the many possibilities will be identical with a naturally occurring molecule. The antibiotic erythromycin, for example, contains 37 carbon, 67 hydrogen, and 13 oxygen atoms, along with one nitrogen atom. Even when joined together in the proper order, these 118 atoms can give rise to 262,144 different structures, only one of which has the characteristics of natural erythromycin. The great abundance of organic compounds, their fundamental role in the chemistry of life, and their structural diversity have made their study especially challenging and exciting. Organic chemistry is the largest area of specialization among the various fields of chemistry.

Biochemistry

As understanding of inanimate chemistry grew during the 19th century, attempts to interpret the physiological processes of living organisms in terms of molecular structure and reactivity gave rise to the discipline of biochemistry. Biochemists employ the techniques and theories of chemistry to probe the molecular basis of life. An organism is investigated on the premise that its physiological processes are the consequence of many thousands of chemical reactions occurring in a highly integrated manner. Biochemists have established, among other things, the principles that underlie energy transfer in cells, the chemical structure of cell membranes, the coding and transmission of hereditary information, muscular and nerve function, and biosynthetic pathways. In fact, related biomolecules have been found to fulfill similar roles in organisms as different as bacteria and human beings. The study of biomolecules, however, presents many difficulties. Such molecules are often very large and exhibit great structural complexity; moreover, the chemical reactions they undergo are usually exceedingly fast. The separation of the two strands of DNA, for instance, occurs in one-millionth of a second. Such rapid rates of reaction are possible only through the intermediary action of biomolecules called enzymes. Enzymes are proteins that owe their remarkable rate-accelerating abilities to their three-dimensional chemical structure. Not surprisingly, biochemical discoveries have had a great impact on the understanding and treatment of disease. Many ailments due to inborn errors of metabolism have been traced to specific genetic defects. Other diseases result from disruptions in normal biochemical pathways.

Frequently, symptoms can be alleviated by drugs, and the discovery, mode of action, and degradation of therapeutic agents is another of the major areas of study in biochemistry. Bacterial infections can be treated with sulfonamides, penicillins, and tetracyclines, and research into viral infections has revealed the effectiveness of acyclovir against the herpes virus. There is much current interest in the details of carcinogenesis and cancer chemotherapy. It is known, for example, that cancer can result when cancer-causing molecules, or carcinogens as they are called, react with nucleic acids and proteins and interfere with their normal modes of action. Researchers have developed tests that can identify molecules likelyto be carcinogenic. The hope, of course, is that progress in the prevention and treatment of cancer will accelerate once the biochemical basis of the disease is more fully understood.

The molecular basis of biologic processes is an essential feature of the fast-growing disciplines of molecular biology and biotechnology. Chemistry has developed methods for rapidly and accurately determining the structure of proteins and DNA. In addition, efficient laboratory methods for the synthesis of genes are being devised. Ultimately, the correction of genetic diseases by replacement of defective genes with normal ones may become possible.

Polymer Chemistry

The simple substance ethylene is a gas composed of molecules with the formula CH_2CH_2. Under certain conditions, many ethylene molecules will join together to form a long chain called polyethylene, with the formula $(CH_2CH_2)_n$, where n is a variable but large number. Polyethylene is a tough, durable solid material quite different from ethylene. It is an example of a polymer, which is a large molecule made up of many smaller molecules (monomers), usually joined together in a linear fashion. Many naturally occurring substances, including cellulose, starch, cotton, wool, rubber, leather, proteins, and DNA, are polymers. Polyethylene, nylon, and acrylics are examples of synthetic polymers. The study of such materials lies within the domain of polymer chemistry, a specialty that has flourished in the 20th century. The investigation of natural polymers overlaps considerably with biochemistry, but the synthesis of new polymers, the investigation of polymerization processes, and the characterization of the structure and properties of polymeric materials all pose unique problems for polymer chemists.

Polymer chemists have designed and synthesized polymers that vary in hardness, flexibility, softening temperature, solubility in water, and biodegradability. They have produced polymeric materials that are as strong as steel yet lighter and more resistant to corrosion. Oil, natural gas, and water pipelines are now routinely constructed of plastic pipe. In recent years, automakers have increased their use of plastic components to build lighter vehicles that consume less fuel. Other industries such as those involved in the manufacture of textiles, rubber, paper, and packaging materials are built upon polymer chemistry.

Besides producing new kinds of polymeric materials, researchers are concerned with developing special catalysts that are required by the large-scale industrial synthesis of commercial polymers. Without such catalysts, the polymerization process would be very slow in certain cases.

Physical Chemistry

Many chemical disciplines, focus on certain classes of materials that share common structural and chemical features. Other specialties may be centred not on a class of substances but rather on their interactions and transformations. The oldest of these fields is physical chemistry, which seeks to measure, correlate, and explain the quantitative aspects of chemical processes. The Anglo-Irish chemist Robert Boyle, for example, discovered in the 17th century that at room temperature the volume of a fixed quantity of gas decreases proportionally as the pressure on it increases. Thus, for a gas at constant temperature, the product of its volume V and pressure P equals a constant number—i.e., $PV = \text{constant}$. Such a simple arithmetic relationship is valid for nearly all gases at room temperature and at pressures equal to or less than one atmosphere. Subsequent work has shown that the relationship loses its validity at higher pressures, but more complicated expressions that more accurately match experimental results can be derived. The discovery and investigation of such chemical regularities, often called laws of nature, lie within the realm of physical chemistry. For much of the 18th century the source of mathematical regularity in chemical systems was assumed to be the continuum of forces and fields that surround the atoms making up chemical elements and compounds. Developments in the 20th century, however, have shown that chemical behaviour is best interpreted by a quantum mechanical model of atomic and molecular structure. The branch of physical chemistry that is largely devoted to this subject is theoretical chemistry. Theoretical chemists make extensive use of computers to help them solve complicated mathematical equations. Other branches of physical chemistry include chemical thermodynamics, which deals with the relationship between heat and other forms of chemical energy, and chemical kinetics, which seeks to measure and understand the rates of chemical reactions. Electrochemistry investigates the interrelationship of electric current and chemical change. The passage of an electric current through a chemical solution causes changes in the constituent substances that are often reversible—i.e., under different conditions the altered substances themselves will yield an electric current. Common batteries contain chemical substances that, when placed in contact with each other by closing an electrical circuit, will deliver current at a constant voltage until the substances are consumed. At present there is much interest in devices that can use the energy in sunlight to drive chemical reactions whose products are capable of storing the energy. The discovery of such devices would make possible the widespread utilization of solar energy.

There are many other disciplines within physical chemistry that are concerned more with the general properties of substances and the interactions among substances than with the substances themselves. Photochemistry is a specialty that investigates the interaction of light with matter. Chemical reactions initiated by the absorption of light can be very different from those that occur by other means. Vitamin D, for example, is formed in the human body when the steroid ergosterol absorbs solar radiation; ergosterol does not change to vitamin D in the dark.

A rapidly developing subdiscipline of physical chemistry is surface chemistry. It examines the properties of chemical surfaces, relying heavily on instruments that can provide a chemical profile of such surfaces. Whenever a solid is exposed to a liquid or a gas, a reaction occurs initially on the surface of the solid, and its properties can change dramatically as a result. Aluminum is a case in point: it is resistant to corrosion precisely because the surface of the pure metal reacts with oxygen to form a layer of aluminum oxide, which serves to protect the interior of the metal from further oxidation. Numerous reaction catalysts perform their function by providing a reactive surface on which substances can react.

Industrial Chemistry

The manufacture, sale, and distribution of chemical products is one of the cornerstones of a developed country. Chemists play an important role in the manufacture, inspection, and safe handling of chemical products, as well as in product development and general management. The manufacture of basic chemicals such as oxygen, chlorine, ammonia, and sulfuric acid provides the raw materials for industries producing textiles, agricultural products, metals, paints, and pulp and paper. Specialty chemicals are produced in smaller amounts for industries involved with such products as pharmaceuticals, foodstuffs, packaging, detergents, flavours, and fragrances. To a large extent, the chemical industry takes the products and reactions common to "bench-top" chemical processes and scales them up to industrial quantities.

The monitoring and control of bulk chemical processes, especially with regard to heat transfer, pose problems usually tackled by chemists and chemical engineers. The disposal of by-products also is a major problem for bulk chemical producers. These and other challenges of industrial chemistry set it apart from the more purely intellectual disciplines of chemistry. Yet, within the chemical industry, there is a considerable amount of fundamental research undertaken within traditional specialties. Most large chemical companies have research-and-development capability. Pharmaceutical firms, for example, operate large research laboratories in which chemists test molecules for pharmacological activity. The new products and processes that are discovered in such laboratories are often patented and become a source of profit for the company funding the research. A great deal of the research conducted in the chemical industry can be termed applied research because its goals are closely tied to the products and processes of the company concerned. New technologies often require much chemical expertise. The fabrication of, say, electronic microcircuits involves close to 100 separate chemical steps from start to finish. Thus, the chemical industry evolves with the technological advances of the modern world and at the same time often contributes to the rate of progress.

Atoms and Molecules

Molecule is a group of two or more atoms that form the smallest identifiable unit into which a pure substance can be divided and still retain the composition and chemical properties of that substance.

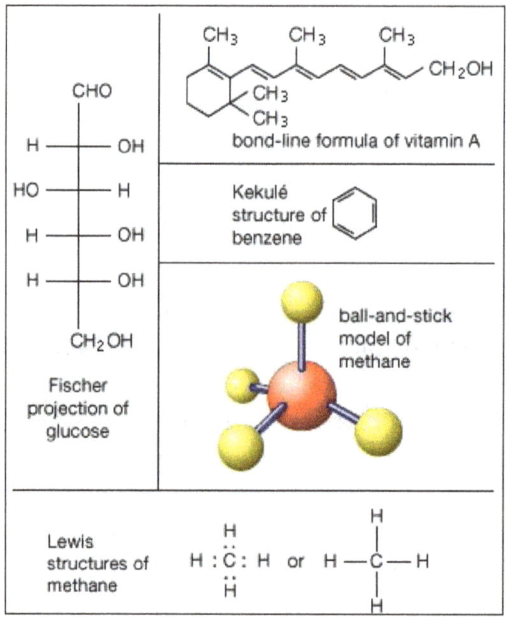

Several methods of representing a molecule's structure. In Lewis structures, element symbols represent atoms, and dots represent electrons surrounding them. A pair of shared electrons (covalent bond) may also be shown as a single dash. The ball-and-stick model better illustrates the spatial arrangement of the atoms. For aromatic compounds, the Kekulé structure is common, in which each bond is represented by a dash, carbon atoms are implied where two or more lines meet, and hydrogen atoms are usually omitted. Bond-line formulas, similar to the Kekulé structure, are often used for complex nonaromatic organic compounds. Sugars are often drawn as Fischer projections, in which the carbon "backbone" is drawn as a straight vertical line, with carbon atoms implied where horizontal lines intersect the vertical one.

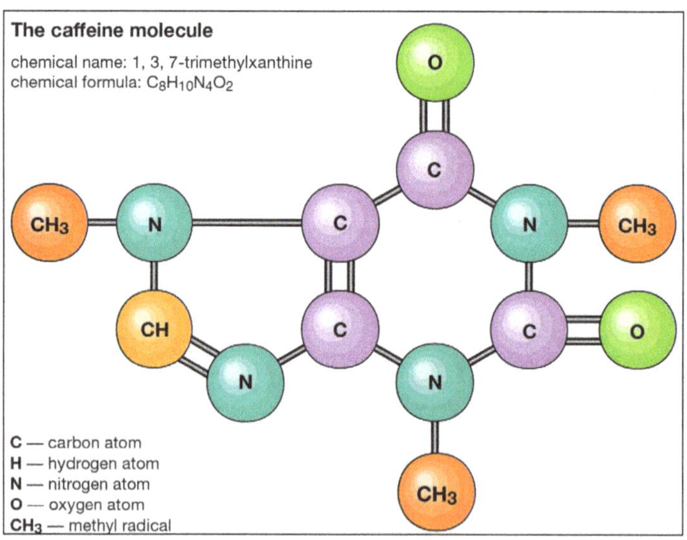

Caffeine Molecule

The division of a sample of a substance into progressively smaller parts produces no change in either its composition or its chemical properties until parts consisting of single molecules are reached. Further subdivision of the substance leads to still smaller parts that usually differ from the original substance in composition and always differ from it in chemical properties. In this latter stage of fragmentation the chemical bonds that hold the atoms together in the molecule are broken.

Atoms consist of a single nucleus with a positive charge surrounded by a cloud of negatively charged electrons. When atoms approach one another closely, the electron clouds interact with each other and with the nuclei. If this interaction is such that the total energy of the system is lowered, then the atoms are bond together to form a molecule. Thus, from a structural point of view, a molecule consists of an aggregation of atoms held together by valence forces. Diatomic molecules contain two atoms that are chemically bonded. If the two atoms are identical, as in, for example, the oxygen molecule (O_2), they compose a homonuclear diatomic molecule, while if the atoms are different, as in the carbon monoxide molecule (CO), they make up a heteronuclear diatomic molecule. Molecules containing more than two atoms are termed polyatomic molecules, e.g., carbon dioxide (CO_2) and water (H_2O). Polymer molecules may contain many thousands of component atoms.

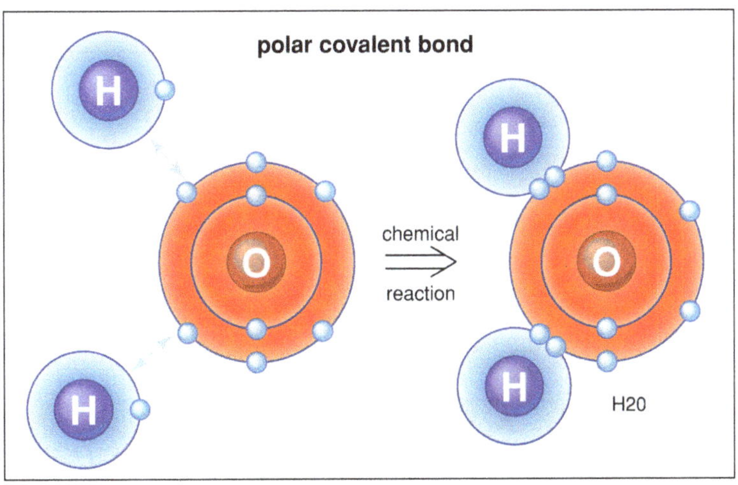

Water Molecule

A water molecule is made up of two hydrogen atoms and one oxygen atom. A single oxygen atom contains six electrons in its outer shell, which can hold a total of eight electrons. When two hydrogen atoms are bound to an oxygen atom, the outer electron shell of oxygen is filled.

The ratio of the numbers of atoms that can be bonded together to form molecules is fixed; for example, every water molecule contains two atoms of hydrogen and one atom of oxygen. It is this feature that distinguishes chemical compounds from solutions and other mechanical mixtures. Thus hydrogen and oxygen may be present in any arbitrary proportions in mechanical mixtures but when sparked will combine only in definite proportions to form the chemical compound water (H_2O). It is possible for the same kinds of atoms to combine in different but definite proportions to form different molecules; for example, two atoms of hydrogen will chemically bond with one atom of oxygen to yield a water molecule, whereas two atoms of hydrogen can chemically bond with two atoms of oxygen to form a molecule of hydrogen peroxide (H_2O_2). Furthermore, it is possible for atoms to bond together in identical proportions to form different molecules. Such molecules are called isomers and differ only in the arrangement of the atoms within the molecules. For example, ethyl alcohol (CH_3CH_2OH) and methyl ether (CH_3OCH_3) both contain one, two, and six atoms of oxygen, carbon, and hydrogen, respectively, but these atoms are bonded in different ways.

Not all substances are made up of distinct molecular units. Sodium chloride (common table salt), for example, consists of sodium ions and chlorine ions arranged in a lattice so that each sodium ion is surrounded by six equidistant chlorine ions and each chlorine ion is surrounded by six equidistant sodium ions. The forces acting between any sodium and any adjacent chlorine ion are equal. Hence, no distinct aggregate identifiable as a molecule of sodium chloride exists. Consequently, in sodium chloride and in all solids of similar type, the concept of the chemical molecule has no significance. Therefore, the formula for such a compound is given as the simplest ratio of the atoms, called a formula unit—in the case of sodium chloride, NaCl.

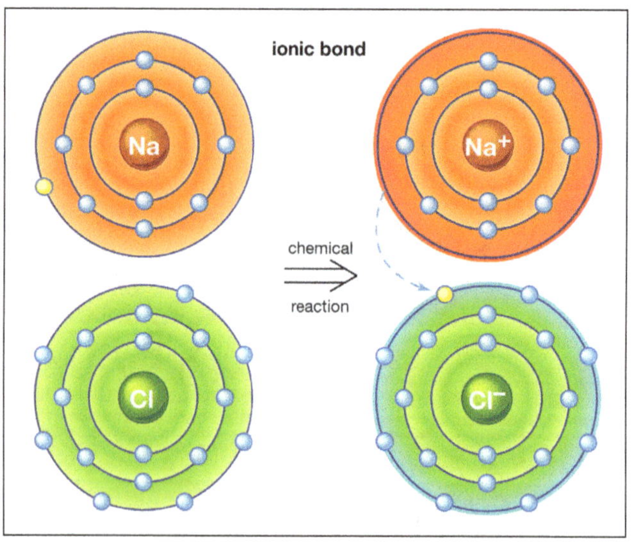

Ionic bond: Sodium chloride or table salt.

Molecules are held together by shared electron pairs, or covalent bonds. Such bonds are directional, meaning that the atoms adopt specific positions relative to one another so as to maximize the bond strengths. As a result, each molecule has a definite, fairly rigid structure, or spatial distribution of its atoms. Structural chemistry is concerned with valence, which determines how atoms combine in definite ratios and how this is related to the bond directions and bond lengths. The properties of molecules correlate with their structures; for example, the water molecule is bent structurally and therefore has a dipole moment, whereas the carbon dioxide molecule is linear and has no dipole moment. The elucidation of the manner in which atoms are reorganized in the course of chemical reactions is important. In some molecules the structure may not be rigid; for example, in ethane (H_3CCH_3) there is virtually free rotation about the carbon-carbon single bond.

Ionic bonding in sodium chloride. An atom of sodium (Na) donates one of its electrons to an atom of chlorine (Cl) in a chemical reaction, and the resulting positive ion (Na^+) and negative ion (Cl^-) form a stable ionic compound (sodium chloride; common table salt) based on this ionic bond.

The nuclear positions in a molecule are determined either from microwave vibration-rotation spectra or by neutron diffraction. The electron cloud surrounding the nuclei in a molecule can be studied by X-ray diffraction experiments. Further information can be obtained by electron spin resonance or nuclear magnetic resonance techniques. Advances in electron microscopy have enabled visual images of individual molecules and atoms to be produced. Theoretically the molecular structure is determined by solving the quantum mechanical equation for the motion of the electrons in the field of the nuclei (called the Schrödinger equation). In a molecular structure the bond lengths and bond angles are those for which the molecular energy is the least. The determination of structures by numerical solution of the Schrödinger equation has become a highly developed process entailing use of computers and supercomputers.

The molecular weight of a molecule is the sum of the atomic weights of its component atoms. If a substance has molecular weight M, then M grams of the substance is termed one mole. The number of molecules in one mole is the same for all substances; this number is known as Avogadro's number ($6.022140857 \times 10^{23}$). Molecular weights can be determined by mass spectrometry and by techniques based on thermodynamics or kinetic transport phenomena.

Elements and Compounds

A chemical element is a pure substance that consists of one type of atom. Each atom has an atomic number, which represents the number of protons that are in the nucleus of a single atom of that element. The periodic table of elements is ordered by ascending atomic number.

The chemical elements are divided into the metals, the metalloids, and the non-metals. Metals, typically found on the left side of the periodic table are:

- Often conductive to electricity.
- Malleable.

- Shiny.
- Sometimes magnetic.

Aluminium, iron, copper, gold, mercury and lead are metals. In contrast, non-metals, found on the right side of the periodic table (to the right of the staircase) are:

- Typically not conductive.
- Not malleable.
- Dull (not shiny).
- Not magnetic.

Examples of elemental non-metals include carbon and oxygen. Metalloids have some characteristics of metals and some characteristics of non-metals. Silicon and arsenic are metalloids.

As of November, 2011, 118 elements have been identified. Of these 118 known elements, only the first 98 are known to occur naturally on Earth. The elements that do not occur naturally on Earth are the synthetic products of man-made nuclear reactions. 80 of the 98 naturally-occurring elements are stable; the rest are radioactive, which means they decay into lighter elements over timescales ranging from fractions of a second to billions of years.

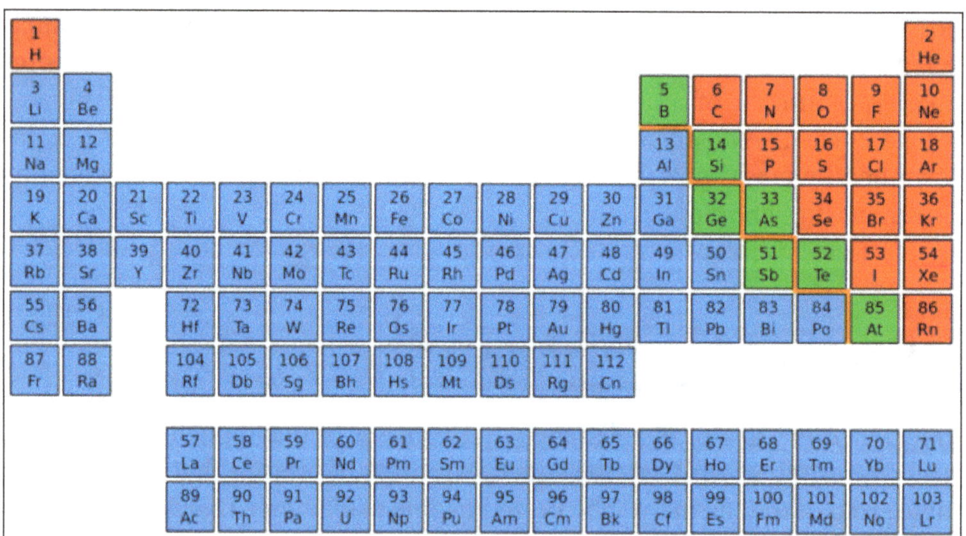

The periodic table.

The periodic table shows 118 elements, including metals (blue), nonmetals (red), and metalloids (green).

Hydrogen and helium are by far the most abundant elements in the universe. However, iron is the most abundant element (by mass) in the composition of the

Earth, and oxygen is the most common element in the layer that is the Earth's crust.

Although all known chemical matter is composed of these elements, chemical matter itself constitutes only about 15% of the matter in the universe. The remainder is dark matter, a mysterious substance that is not composed of chemical elements. Dark matter lacks protons, neutrons, or electrons.

Compounds

Pure samples of isolated elements are uncommon in nature. While the 98 naturally occurring elements have all been identified in mineral samples from the Earth's crust, only a small minority of them can be found as recognizable, relatively pure minerals. Among the more common of such "native elements" are copper, silver, gold, and sulfur. Carbon is also commonly found in the form of coal, graphite, and diamonds. The noble gases (e.g., neon) and noble metals (e.g., mercury) can also be found in their pure, non-bonded forms in nature. Still, most of these elements are found in mixtures.

When two distinct elements are chemically combined—i.e., chemical bonds form between their atoms—the result is called a chemical compound. Most elements on Earth bond with other elements to form chemical compounds, such as sodium (Na) and Chloride (Cl), which combine to form table salt (NaCl). Water is another example of a chemical compound. The two or more component elements of a compound can be separated through chemical reactions.

Chemical compounds have a unique and defined structure, which consists of a fixed ratio of atoms held together in a defined spatial arrangement by chemical bonds. Chemical compounds can be:

- Molecular compounds held together by covalent bonds.
- Salts held together by ionic bonds.
- Intermetallic compounds held together by metallic bonds.
- Complexes held together by coordinate covalent bonds.

Pure chemical elements are not considered chemical compounds, even if they consist of diatomic or polyatomic molecules (molecules that contain only multiple atoms of a single element, such as H_2 or S_8).

Substance and Mixture

In chemistry, a chemical substance is a form of matter that has constant chemical composition and characteristic properties. It cannot be separated into components without breaking chemical bonds. Chemical substances can be solids, liquids, gases, or plasma. Changes in temperature or pressure can cause substances to shift between the different phases of matter. An element is a chemical substance that is made up of a particular kind of atom and hence cannot be broken down or transformed by a chemical reaction into a different element. All atoms of an element have the same number of protons, though they may have different numbers of neutrons and electrons.

A pure chemical compound is a chemical substance that is composed of a particular set of molecules or ions that are chemically bonded. Two or more elements combined into one substance through a chemical reaction, such as water, form a chemical compound. All compounds are substances, but not all substances are compounds. A chemical compound can be either atoms bonded together in molecules or crystals in which atoms, molecules or ions form a crystalline lattice. Compounds made primarily of carbon and hydrogen atoms are called organic compounds, and all others are called inorganic compounds. Compounds containing bonds between carbon and a metal are called organometallic compounds.

Chemical substances are often called 'pure' to set them apart from mixtures. A common example of a chemical substance is pure water; it always has the same properties and the same ratio of hydrogen to oxygen whether it is isolated from a river or made in a laboratory. Other chemical substances commonly encountered in pure form are diamond (carbon), gold, table salt (sodium chloride), and refined sugar (sucrose). Simple or seemingly pure substances found in nature can in fact be mixtures of chemical substances. For example, tap water may contain small amounts of dissolved sodium chloride and compounds containing iron, calcium, and many other chemical substances. Pure distilled water is a substance, but seawater, since it contains ions and complex molecules, is a mixture.

Chemical Mixtures

Naturally occurring sulfur crystals.

A mixture is a material system made up of two or more different substances, which are mixed but not combined chemically. A mixture refers to the physical combination of two or more substances in which the identities of the individual substances are retained. Mixtures take the form of alloys, solutions, suspensions, and colloids.

Sulfur occurs naturally as elemental sulfur, sulfide, and sulfate minerals and in hydrogen sulfide. This mineral deposit is composed of a mixture of substances.

Heterogeneous Mixtures

A heterogeneous mixture is a mixture of two or more chemical substances (elements or compounds), where the different components can be visually distinguished and easily separated by physical means. Examples include:

- Mixtures of sand and water.
- Mixtures of sand and iron filings.
- A conglomerate rock.
- Water and oil.
- A salad.
- Trail mix.
- Mixtures of gold powder and silver powder.

Homogenous Mixtures

A homogeneous mixture is a mixture of two or more chemical substances (elements or compounds), where the different components cannot be visually distinguished. The composition of homogeneous mixtures is constant. Often separating the components of a homogeneous mixture is more challenging than separating the components of a heterogeneous mixture.

Distinguishing between homogeneous and heterogeneous mixtures is a matter of the scale of sampling. On a small enough scale, any mixture can be said to be heterogeneous, because a sample could be as small as a single molecule. In practical terms, if the property of interest is the same regardless of how much of the mixture is taken, the mixture is homogeneous. A mixture's physical properties, such as its melting point, may differ from those of its individual components. Some mixtures can be separated into their components by physical (mechanical or thermal) means.

Phase

States: Gas, Liquid and Solid

Gas, liquid, and solid are known as the three states of matter or material, but each of solid and liquid states may exist in one or more forms. Thus, another term is required to describe the various forms, and the term phase is used. Each distinct form is called a phase; however, the concept of phase defined as a homogeneous portion of a system extends beyond a single material, because a phase may also involve several materials. For example, a homogeneous solution of any number of substances is a one-phase system. Phase is a concept used to explain many physical and chemical changes (reactions).

- A solid has a definite shape and volume. A liquid has a definite volume but it takes the shape of a container whereas a gas fills the entire volume of a container. You already know that diamond and graphite are solids made up of the element carbon; they are two phases of carbon, but both are solids. Solids are divided into subclasses of amorphous (or glassy) solids and crystalline solids. Arrangements of atoms or molecules in crystalline solids are repeated regularly over a very long range of millions of atoms, but their arrangements in amorphous solids are somewhat random or short range of say some tens or hundreds of atoms.
- In general, there is only one liquid phase of a material. However, there are two forms of liquid helium; each have some unique properties. Thus, the two forms are different (liquid) phases of helium. At a definite temperature and pressure, the two phases co-exist. So far, all gases behave alike as do mixtures of gases. Thus, a gas is usually considered as a phase.

The Concept of Phase

A phase is a distinct and homogeneous state of a system with no visible boundary separating it into parts. Water, H_2O, is such a common substance that its gas (steam), liquid (water), and solid (ice) phases are widely known. An ice water mixture has two phases, as do systems containing ice-and-vapor, and water-and-vapor. To recognize the vapor system in these systems may require a keen observation, because the vapor usually blends with air, and is not detected directly.

You probably also know that several solids may exist for a substance, and each of the solid forms is also called a phase. Diamond and graphite are the most quoted examples; both are solid carbon, but they have different crystal shapes, colors, and structures. They represent two different phases of carbon. Ice is another example, under 1 atm, ice has hexagonal symmetry, while cubic ice is formed under high pressure. In fact, there are at least eight different types of ice, each being a solid phase.

When you mix water and alcohol, regardless of the relative amounts that you use, they are fully miscible. The resulting mixture has only one phase (a solution). However, water and oil are normally immiscible, and their boundary of separation is visible; they form a two-phase system. Sometimes you cannot "see" the boundary, and you will need scientific reasoning to realize the number of phases present in system.

Well, there are so many concepts packed into one term that we cannot make the definition any simpler for you. However, the term is useful because it can be used to explain many phenomena. There is no substitute for it.

Phase Transitions

The conversion between these phases is called a phase transition. A state change of any material due to temperature or pressure change is a phase transition. A phase transition is a physical change (or reaction). The following diagram illustrates the key phase transitions: You should know the names of the process for these phase transitions.

$$\text{SOLID} \xrightarrow{\text{sublimation}} \text{GAS} \xrightarrow{\text{deposition}} \text{SOLID}$$

$$\text{SOLID} \xrightarrow{\text{melting}} \text{LIQUID} \xrightarrow[\text{(solidfiction)}]{\text{freezing}} \text{SOLID}$$

$$\text{GAS} \xrightarrow{\text{condensation}} \text{LIQUID} \xrightarrow{\text{vaporization}} \text{GAS}$$

The concepts of phase and phase transition introduce you into fields of materials studies. For example, if you search the internet with the phrase "phase transition", you get thousands of websites; some are related to the concept we have discussed here, but some may be using "phase transition" as a catchy phrase. The concept of phase transition is also applied to the study of nuclear matter such as protons and neutrons.

Bonding

In chemistry, a bond or chemical bond is a link between atoms in molecules or compounds and between ions and molecules in crystals. A bond represents a lasting attraction between different atoms, molecules or ions.

Why Bonds Form

Most of bonding behaviour can be explained by the attraction between two opposite electrical charge. The electrons of an atom or ion are attracted to their own positively-charged nucleus (containing protons), yet also to the nuclei of nearby atoms. Species that participate in chemical bonds are more stable when the bond is formed, typically because they had an imbalance of charge (greater or fewer number of electrons than protons) or because their valence electrons did not fill or half-fill electron orbitals.

Examples of Chemical Bonds

The two main types of bonds are covalent bonds and ionic bonds. Covalent bonding is where atoms share electrons more or less equally between each other. In an ionic bond, an electron from one atom spends more time associated with the nucleus and electron orbitals of the other atom (essentially donated). However, pure covalent and ionic bonding is relatively rare. Usually a bond is intermediate between ionic and covalent. In a polar covalent bond, electrons are shared, but the electrons participating in the bond are more attracted to one atom than to the other.

Another type of bonding is a metallic bond. In a metallic bond, electrons are donated to an "electron sea" between a group of atoms. Metallic bonding is very strong, but the fluid nature of the electrons allows for a high degree of electrical and thermal conductivity.

Energy

Energy is a property of objects which can be transferred to other objects or converted into different forms, but cannot be created or destroyed. Organisms use energy to survive, grow, respond to stimuli, reproduce, and for every type of biological process. The potential energy stored in molecules can be converted to chemical energy, which can ultimately be converted to kinetic energy, enabling an organism to move. Eventually, most of energy used by organisms is transformed into heat and dissipated.

Kinetic Energy

Energy associated with objects in motion is called kinetic energy. For example, when an airplane is in flight, the airplane is moving through air very quickly—doing work to enact change on its surroundings. The jet engines are converting potential energy in fuel to the kinetic energy of movement. A wrecking ball can perform a large amount of damage, even when moving slowly. However, a still wrecking ball cannot perform any work and therefore has no kinetic energy. A speeding bullet, a walking person, the rapid movement of molecules in the air that produces heat, and electromagnetic radiation, such as sunlight, all have kinetic energy.

Potential Energy

What if that same motionless wrecking ball is lifted two stories above a car with a crane? If the suspended wrecking ball is not moving, is there energy associated with it? Yes, the wrecking ball has energy because the wrecking ball has the potential to do work. This form of energy is called potential energy because it is possible for that object to do work in a given state.

Objects transfer their energy between potential and kinetic states. As the wrecking ball hangs motionlessly, it has [latex]\text{0%}[/latex] kinetic and [latex]\text{100%}[/latex] potential energy. Once the ball is released, its kinetic energy increases as the ball picks up speed. At the same time, the ball loses potential energy as it nears the ground. Other examples of potential energy include the energy of water held behind a dam or a person about to skydive out of an airplane.

Potential energy vs. kinetic energy: Water behind a dam has potential energy. Moving water, such as in a waterfall or a rapidly flowing river, has kinetic energy.

Chemical Energy

Chemical energy: The molecules in gasoline (octane, the chemical formula shown) contain chemical energy. This energy is transformed into kinetic energy that allows a car to race on a racetrack.

Potential energy is not only associated with the location of matter, but also with the structure of matter. A spring on the ground has potential energy if it is compressed, as does a rubber band that is pulled taut. The same principle applies to molecules. On a chemical level, the bonds that hold the atoms of molecules together have potential energy. This type of potential energy is called chemical energy, and like all potential energy, it can be used to do work.

For example, chemical energy is contained in the gasoline molecules that are used to power cars. When gas ignites in the engine, the bonds within its molecules are broken, and the energy released is used to drive the pistons. The potential energy stored within chemical bonds can be harnessed to perform work for biological processes. Different metabolic processes break down organic molecules to release the energy for an organism to grow and survive.

Reaction

A chemical reaction is in which the bonds are broken within reactant molecules, and new bonds are formed within product molecules in order to form a new substance. Chemical reactions are all around us, from the metabolism of food in our body to how the light we get from the sun is the results of chemical reactions. Before beginning with chemical reactions, it is important to know about physical and chemical changes.

A burning candle is the best example of physical and chemical change. Take a candle and light it. As time passes, we can observe that the candle changes to wax. If you cover the candle with a jar, it will extinguish.

In the demonstration, burning of the candle is a chemical change while conversion of the candle to wax is a physical change. In a physical change, there is basically a change of state of the substance but in the case of a chemical change mostly a new substance is formed in which either energy is given off or absorbed. Thus, we can conclude that chemical changes are accompanied by certain physical changes.

Basic Concepts of Chemical Reactions

- A Chemical Reaction is a process that occurs when two or more molecules interact to form a new product(s).

- Compounds that interact to produce new compounds are called reactants whereas the newly formed compounds are called products.
- Chemical reactions play an integral role in different industries, customs and even in our daily life. They are continuously happening in our general surroundings; for example, rusting of iron, pottery, fermentation of wine and so on.
- In a chemical reaction, a chemical change must occur which is generally observed with physical changes like precipitation, heat production, colour change etc.
- A reaction can take place between two atoms or ions or molecules, and they form a new bond and no atom is destroyed or created but a new product is formed from reactants.
- The rate of reaction depends on and is affected by factors like pressure, temperature, the concentration of reactants.

Chemical Equations

Due to the vast amounts of chemical reactions happening around us, a nomenclature was developed to simplify how we express a chemical reaction in the form of a chemical equation. A chemical equation is nothing but a mathematical statement which symbolizes the product formation from reactants while stating certain condition for which how the reaction has been conducted.

The reactants are on the left-hand side whereas products formed on the right-hand side connected by a one-headed or two-headed arrows. For example, a reaction:

$$A + B \rightarrow C + D$$

Here, A and B are the reactants, which react to form the products C and D. In an actual chemical equation, reactants are denoted by their chemical formula. In order to assure the law of conservation of mass, a chemical equation must be balanced i.e. the number of atoms on both sides must be equal. This is the balancing of the equation.

Let us consider an actual chemical reaction between Methane(CH_4) and Oxygen (O_2),

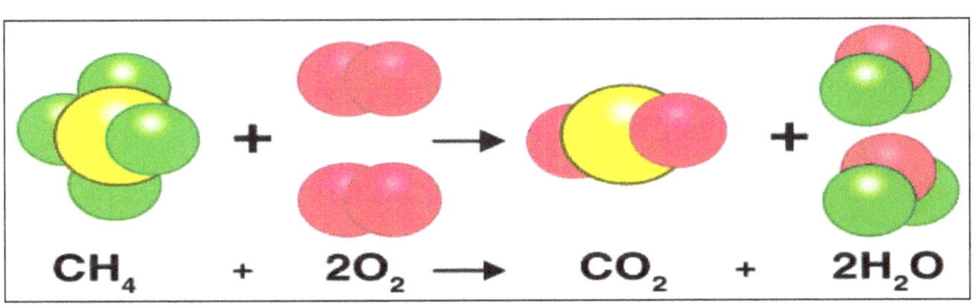

Combustion chemical reaction.

Here we can see how the number of each atom on the left side is balanced on the right side, as stated by the law of conservation of mass.

Types of Chemical Reactions

The basis for different types of reactions is the product formed, the changes that occur, the reactants involved and so on. Different types of reactions are:

- Combustion reaction.
- Decomposition reaction.
- Neutralization reaction.
- Redox Reaction.
- Precipitation or Double-Displacement Reaction.
- Synthesis reaction.

Combustion Reaction

A combustion reaction is a reaction with a combustible material with an oxidizer to give an oxidized product. An oxidizer is a chemical a fuel requires to burn, generally oxygen. Consider the example of combustion of magnesium metal.

$$2Mg + O_2 \rightarrow 2MgO + Heat$$

Here, 2 magnesium atoms react with a molecule of oxygen producing 2 molecules of the compound magnesium oxide releasing some heat in the process.

Decomposition Reaction

A Decomposition reaction is a reaction in which a single component breaks down into multiple products. Certain changes in energy in the environment have to be made like heat, light or electricity breaking bonds of the compound. Consider the example of the decomposition of calcium carbonate giving out CaO (Quick Lime) which is a major component of cement.

$$CaCO_3(s) \xrightarrow{Heat} CaO(s) + CO_2(g)$$

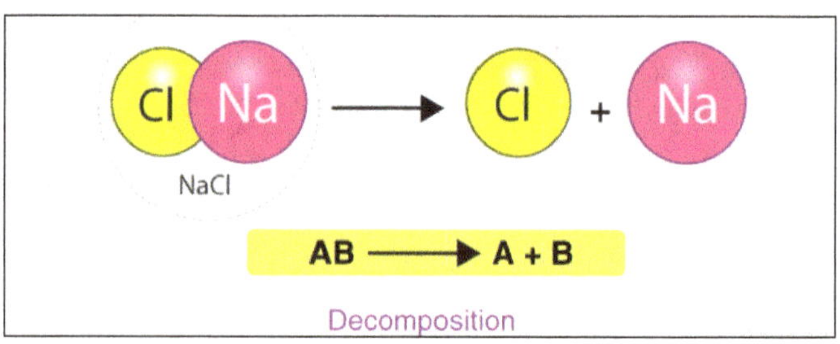

Here, the compound Calcium carbonate when heated breaks down into Calcium Oxide and Carbon Dioxide.

Neutralization Reaction

A Neutralization reaction is basically the reaction between an acid and a base giving salt and water as the products. The water molecule formed is by the combination of OH^- ions and H^+ ions. The overall pH of the products when a strong acid and a strong base undergo a neutralization reaction will be 7. Consider the example of the neutralization reaction between Hydrochloric acid and Sodium Hydroxide giving out sodium chloride(Common Salt) and water.

$$HCl + NaOH \rightarrow NaCl + H_2O$$

Here, an acid and a base, Hydrochloric acid and Sodium Hydroxide react in a neutralization reaction to produce Sodium Chloride(Common Salt) and water as the products.

Redox Reaction

A reduction-oxidation reaction is a reaction in which there is a transfer of electrons between chemical species. Let us consider the example of an electrochemical cell-like redox reaction between Zinc and Hydrogen.

$$Zn + 2H^+ \rightarrow Zn^{2+} + H_2$$

Here, A Zinc atom reacts with 2 ions of positively charged hydrogen to which electrons get transferred from zinc atom and hydrogen becomes a stable molecule and Zinc ion is the product.

Precipitation or Double-Displacement Reaction

It is a type of displacement reaction in which two compounds react and consequently, their anions and cations switch places forming two new products. Consider the example of the reaction between silver nitrate and sodium chloride. The products will be silver chloride and sodium nitrate after the double-displacement reaction.

$$AgNO_3 + NaCl \rightarrow AgCl + NaNO_3$$

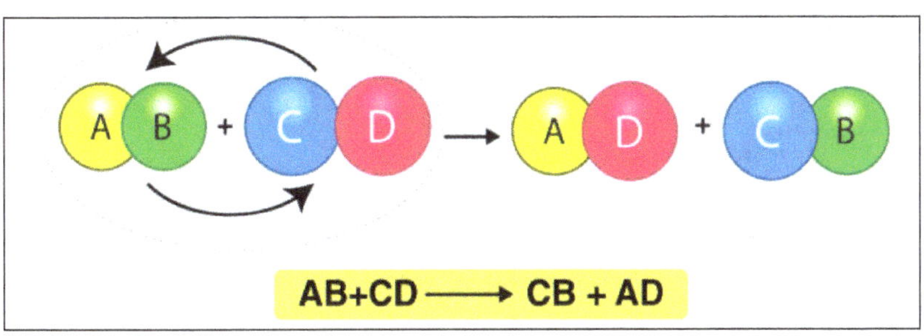

Double decomposition reactions.

Here, Silver Nitrate and Sodium Chloride undergo a double displacement reaction.

Wherein Silver replaces Sodium in Sodium Chloride and Sodium joins with Nitrate becoming Sodium Nitrate along with the Silver Chloride as the product.

Synthesis Reaction

A Synthesis reaction is one of the most basic types of reaction wherein multiple simple compounds combine under certain physical conditions giving out a complex product. The product will always be a compound. Let us consider the Synthesis reaction of sodium chloride with reactants solid sodium and chloride gas.

$$2Na(s)+Cl(g)\rightarrow 2NaCl(s)$$

Here, we have 2 Atoms of solid Sodium reacting with Chlorine gas giving out Sodium Chloride viz. Common Salt as the product:

- In a chemical change, a new compound is formed but in a physical change, the substance changes its state of existence.
- Atoms or ions or molecules which react to form a new substance are called reactants; the new atoms or molecules formed are products.
- A chemical reaction follows the law of conservation of mass. That is no atom is destroyed or created but only a new product is formed from reactants.

Acidity and Basicity

Organic chemists use two acid-base theories for interpreting and planning their work: the Brønsted theory and the Lewis theory.

Brønsted Theory

According to the Brønsted theory, an acid is a proton donor, and a base is a proton acceptor. In an acid-base reaction, each side of the equilibrium has an acid and a base reactant or product, and these may be neutral species or ions.

$$\underset{(acid^1)}{H-A} + \underset{(base^1)}{B:^{(-)}} \rightleftharpoons \underset{(base^2)}{A:^{(-)}} + \underset{(acid^2)}{B-H}$$

Structurally related acid-base pairs, such as {H-A and $A:^{(-)}$} or {$B:^{(-)}$ and B-H} are called conjugate pairs. Substances that can serve as both acids and bases, such as water, are termed amphoteric.

$$\underset{(acid)}{H-Cl} + \underset{(base)}{H_2O} \rightleftharpoons \underset{(base)}{Cl:^{(-)}} + \underset{(acid)}{H_3O^{(+)}}$$

$$\underset{(\text{base})}{H_3N} + \underset{(\text{acid})}{H_2O} \rightleftharpoons \underset{(\text{acid})}{NH_4^{(+)}} + \underset{(\text{base})}{HO^{(-)}}$$

- The stronger the acid the weaker its conjugate base; the stronger the base the weaker its conjugate acid.
- Acid-base equilibria always favor the weakest acid and the weakest base.

Acidity

The relative strength of a group of acids may be evaluated by measuring the extent of reaction that each member undergoes with a common base. Water serves nicely as the common base for such determinations. Thus, for an acid H-A, its strength is proportional to the extent of its reaction with the base water, which is given by the equilibrium constant K_{eq}.

$$H-A+H_2O \rightleftharpoons H_3O^{(+)}+A:^{(-)} \qquad K_{eq}=\frac{[H_3O^{\oplus}][A:^{\ominus}]}{[HA][H_2O]}$$

Since these studies are generally extrapolated to high dilution, the molar concentration of water (55.5) is constant and may be eliminated from the denominator. The resulting K value is called the acidity constant, K_a. Clearly, strong acids have larger K_a's than do weaker acids. Indeed, if HA is completely ionized in solution, K_a approaches infinity ([HA] is nearly zero). Because of the very large range of acid strengths (greater than 10^{40}), a logarithmic scale of acidity (pK_a) is normally employed. Stronger acids have smaller or more negative pK_a values than do weaker acids.

$$K_a=\frac{[H_3O^{\oplus}][A:^{\ominus}]}{[HA]} \qquad pK_a=-\log K_a=\log\left(\frac{1}{K_a}\right)$$

Examples of Brønsted Acid-Base Equilibria

Acid-Base Reaction	Conjugate Acids	Conjugate Bases	K_a	pK_a
$HF+H_2O \rightleftharpoons H_3O^{(+)}+F^{(-)}$	HF $H_3O^{(+)}$	$F^{(-)}$ H_2O	$6.6*10^{-4}$	3.2
$CH_3CO_2H+ \rightleftharpoons H_3O^{(+)}+CH_3CO_2^{(-)}$	CH_3CO_2H $H_3O^{(+)}$	$CH_3CO_2^{(-)}$ H_2O	$1.77*10^{-5}$	4.75

$HOCl + H_2O \rightleftharpoons H_3O^{(+)} + ClO^{(-)}$	HOCl $H_3O^{(+)}$	$ClO^{(-)}$ H_2O	2.95*10⁻⁸	7.53
$H_2O_2 + H_2O \rightleftharpoons H_3O^{(+)} + HO_2(-)$	H_2O_2 $H_3O^{(+)}$	$HO_2^{(-)}$ H_2O	2.5*10⁻¹²	11.6

In all the above examples water acts as a common base, and the weakest acid and the weakest base will predominate, and will necessarily be on the same side of the equilibrium). Idealy, acid concentrations no larger than 0.1M should be used. Since water is amphoteric, it can undergo autoprotolysis as shown in the following equation. The autoionization is shown by Keq, and treating the water concentration as before, the acidity constant (K_a) for water is designated K_w (the water constant). The pH of pure water is therefore 7.0.

$$2H_2O \rightleftharpoons H_3O^{(+)} + OH^{(-)} \qquad K_{eq} = \frac{\left[H_3\overset{\oplus}{O}\right]\left[H\overset{\ominus}{O}\right]}{[H_2O]^2} \qquad K_w = \left[H_3\overset{\oplus}{O}\right] = 10^{-14}$$

The range of strong acids that can be distinguished is limited. Thus acids that are stronger than the hydronium cation, $H_3O^{(+)}$, and weak acids having conjugate bases stronger than hydroxide anion, $OH^{(-)}$, cannot be measured directly in water solution. If a strong acid (HA) has a Ka = 100, a 0.1M solution in water will ionize completely, giving a hydronium concentration of 100 and a [HA] concentration of 10^{-4}. A similar solution of a 10^3 times stronger acid would be undetectably different due to this leveling effect of the solvent. The term leveling effect refers to the influence an amphoteric solvent such as water has on measurements of acidity and basicity. The strength of a strong acid is leveled by the basicity of the solvent, and the basicity of strong bases is likewise limited by solvent acidity.

Since many organic reactions either take place in aqueous environments (living cells), or are quenched or worked-up in water, it is important to consider how a conjugate acid-base equilibrium mixture changes with pH. A simple relationship known as the Henderson-Hasselbalch equation provides this information.

$$\text{Henderson–Hasselbalch Equation}: pK_a = pH + \log\frac{[HA]}{[A^-]}$$

When the pH of an aqueous solution or mixture is equal to the pK_a of an acidic component, the concentrations of the acid and base conjugate forms must be equal (the log of 1 is 0). If the pH is lowered by two or more units relative to the pK_a, the acid concentration will be greater than 99%. On the other hand, if the pH (relative to pK_a) is raised by two or more units the conjugate base concentration will be over 99%. Consequently, mixtures of acidic and non-acidic compounds are easily separated by adjusting the pH of the water component in a two phase solvent extraction.

Basicity

In aqueous solution a base is a solute that increases the concentration of hydroxide ions. Metal alkoxides such as $NaOCH_3$ and $KOC(CH_3)_3$ react with water giving solvated metal cations and alkoxide and hydroxide ions. These 100% ionizations make such compounds very strong bases. Many neutral bases are compounds of nitrogen. For such, we may write base-acid equilibria, which define a K_b and a corresponding pK_b, as shown for ammonia in the first equation below. However, a more common procedure is to report the acidities of the conjugate acids of the bases, as shown in the second equation (these conjugate acids are usually "onium" cations). The pK_a's reported for bases in this system are proportional to the base strength of the base. Note that in such relationships $pK_a + pK_b = 14$.

Acid-Base Reaction	Conjugate Acids	Conjugate Bases	K	pK
$NH_3 + H_2O \rightleftarrows NH_4^{(+)} + OH^{(-)}$	$NH_4^{(+)}$ H_2O	NH_3 $OH^{(-)}$	$K_b = 1.8*10^{-5}$	$pK_b = 4.74$
$NH_4^{(+)} + H_2O \rightleftarrows H_3O^{(+)} + NH_3$	$NH_4^{(+)}$ $H_3O^{(+)}$	NH_3 H_2O	$K_a = 5.5*10^{-10}$	$pK_a = 9.25$

Most simple alkyl amines have pK_a's in the range 9.5 to 11.0, and their water solutions are basic (have a pH of 11 to 12, depending on concentration). The first four compounds in the following table, including ammonia, fall into that category. The last six compounds (colored cells) are significantly weaker bases as a consequence of three factors. The first of these is the hybridization of the nitrogen. In pyridine the nitrogen is sp^2 hybridized, and in nitriles (last entry) an sp hybrid nitrogen is part of the triple bond. In each of these compounds (shaded red) the non-bonding electron pair is localized on the nitrogen atom, but increasing s-character brings it closer to the nitrogen nucleus, reducing its tendency to bond to a proton. Secondly, aniline and p-nitroaniline (first two green shaded structures) are weaker bases due to delocalization of the nitrogen non-bonding electron pair into the aromatic ring (and the nitro substituent). This is the same delocalization that results in activation of a benzene ring toward electrophilic substitution. Indeed, aniline is a weaker base than cyclohexyl amine by roughly a million fold, the same factor by which phenol is a stronger acid than cyclohexanol. This electron pair delocalization is accompanied by a degree of rehybridization of the amino nitrogen atom, but the electron pair delocalization is probably the major factor in the reduced basicity of these compounds. A similar electron pair delocalization is responsible for the very low basicity (and nucleophilic reactivity) of amide nitrogen atoms (last green shaded structure).

Com-pound	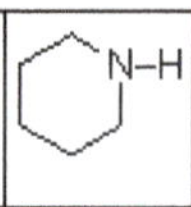	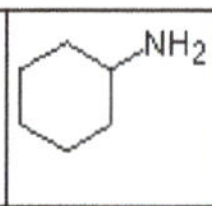	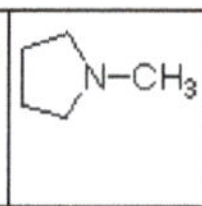	NH_3	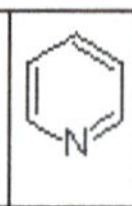	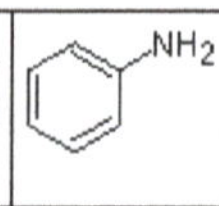	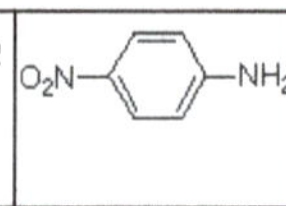	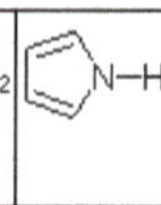	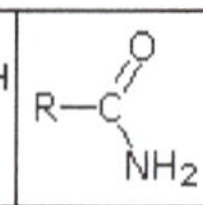	$CH_3C{\equiv}N$
pK_a (onium)	11.0	10.7	10.7	9.3	5.2	4.6	1.0	0.0	-1.0	-10.0

Lewis Theory

According to the Lewis theory, an acid is an electron pair acceptor, and a base is an electron pair donor. Lewis bases are also Brønsted bases; however, many Lewis acids, such as BF_3, $AlCl_3$ and Mg^{2+}, are not Brønsted acids. The product of a Lewis acid-base reaction, is a neutral, dipolar or charged complex, which may be a stable covalent molecule. As shown at the top of the following drawing, coordinate covalent bonding of a phosphorous Lewis base to a boron Lewis acid creates a complex in which the formal charge of boron is negative and that of phosphorous is positive. In this complex, boron acquires a neon valence shell configuration and phosphorous an argon configuration. Two examples of Lewis acid-base equilibria that play a role in chemical reactions are shown in equations 1 & 2 below.

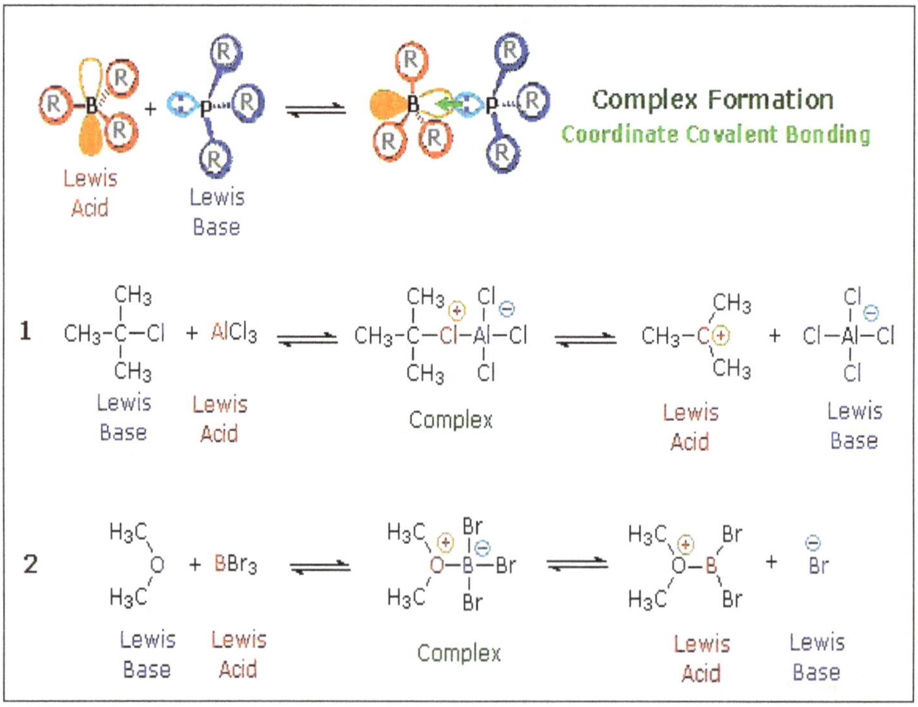

In the first example, an electron deficient aluminum atom bonds to a covalent chlorine atom by sharing one of its non-bonding valence electron pairs, and thus achieves an argon-like valence shell octet. Because this sharing is unilateral (chlorine contributes both electrons), both the aluminum and the chlorine have formal charges, as shown. If the carbon chlorine bond in this complex breaks with both the bonding electrons remaining with the more electronegative atom (chlorine), the carbon assumes a positive charge. We refer to such carbon species as carbocations. Carbocations are also Lewis acids, as the reverse reaction demonstrates. If the substituents (R) on these atoms are not large, the complex will be favored at equilibrium. However, steric hindrance of bulky substituents may prohibit complex formation. The resulting mixture of non-bonded Lewis acid/base pairs has been termed "frustrated", and exhibits unusual chemical and catalytic behavior. The combination of P(t-Butyl)$_3$ and B(C_6F_5)$_3$ is an example.

Many carbocations (but not all) may also function as Brønsted acids. Equation 3 illustrates this dual behavior for $(CH_3)_3C(+)$; the Lewis acidic site is colored red and three of the nine acidic hydrogen atoms are colored orange. In its Brønsted acid role the carbocation donates a proton to the base (hydroxide anion), and is converted to a stable neutral molecule having a carbon-carbon double bond.

3 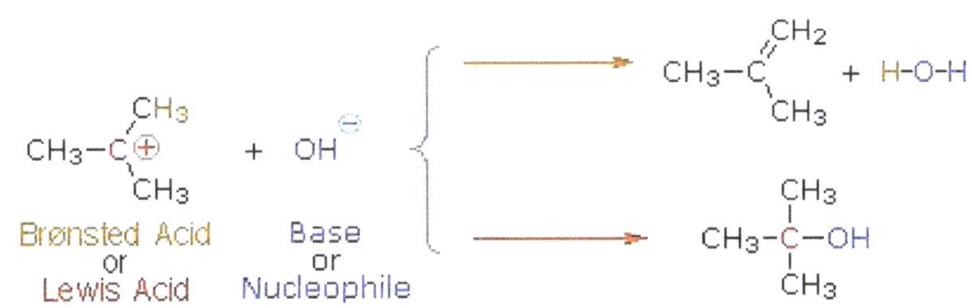

A terminology related to the Lewis acid-base nomenclature is often used by organic chemists. Here the term electrophile corresponds to a Lewis acid, and nucleophile corresponds to a Lewis base.

- Electrophile: An electron deficient atom, ion or molecule that has an affinity for an electron pair, and will bond to a base or nucleophile.
- Nucleophile: An atom, ion or molecule that has an electron pair that may be donated in bonding to an electrophile (or Lewis acid).

Some confusion in distinguishing basicity (base strength) and nucleophilicity (nucleophile strength) is inevitable. Since basicity is a less troublesome concept; it is convenient to start with it. Basicity refers to the ability of a base to accept a proton. Basicity may be related to the pK_a of the corresponding conjugate acid. The strongest bases have the weakest conjugate acids and *vice versa*. In an acid-base equilibrium the weakest acid and the weakest base will predominate and they will necessarily be on the same side of the equilibrium.

Solvents such as acetic acid, acetonitrile and nitromethane are often used for studying very strong acids. Likewise, very weakly acidic solvents such as dimethylsulfoxide (DMSO), acetonitrile, toluene, amines and ammonia may be used to study the acidities of very weak acids. Measurements in these solvents may be extrapolated to water and many have large uncertainties. For both acids and bases, the reported pK_a values extrapolated to water are approximate. A table of some basic reagents is shown below. The range of basicities is remarkable, covering over fifty powers of ten!

Base	$I^{(-)}$	$Cl^{(-)}$	H_2O	$CH_3CO_2^{(-)}$	$RS^{(-)}$	$CN^{(-)}$	$RO^{(-)}$	$NH_2^{(-)}$	$CH_3^{(-)}$
Conj. Acid	HI	HCl	$H_3O^{(+)}$	CH_3CO_2H	RSH	HCN	ROH	NH_3	CH_4
pK_a	-9	-7	-1.7	4.8	8	9.1	16	33	48

Increasing base strength →

Nucleophilicity is a more complex property. It commonly refers to the rate of substitution reactions at the halogen-bearing carbon atom of a reference alkyl halide, such as CH_3-Br. Thus the nucleophilicity of the $Nu:^{(-)}$reactant in the following substitution reaction varies as shown in the chart below. The reactivity range encompassed by these reagents is over 5,000 fold, thiolate being the most reactive. Clearly, there are significant differences between these nucleophilicities and the basicities.

$$CH_3-Br+\overset{\ominus}{Nu}: \xrightarrow{CH_3OH\ soln} CH_3-Nu+\overset{\ominus}{Br}:$$

$$\text{Nucleophilicity: } CH_3CO_2^{(-)} < Cl^{(-)} < Br^{(-)} < N_3^{(-)} < CH_3O^{(-)} < CN^{(-)} < I^{(-)} < CH_3S^{(-)}$$

Increasing nucleophile strength →

Some useful trends have been documented:

- For a given element, negatively charged species are more nucleophilic (and basic) than are equivalent neutral species.
- For a given period of the periodic table, nucleophilicity (and basicity) decreases on moving from left to right.
- For a given group of the periodic table, nucleophilicity increases from top to bottom (i.e. with increasing size), although there is a solvent dependence due to hydrogen bonding. Basicity varies in the opposite manner.

For two or more molecules incorporating nucleophilic atoms of the same kind and charge, the stronger base is usually the stronger nucleophile. Thus, 2, 2, 2-trifluroethoxide (pK_a 12) is a weaker base and nucleophile than ethoxide (pK_a 16). A notable exception to this rule occurs when a vicinal (adjacent) atom carries a non-bonding electron pair. Two common examples of this exception, called the alpha effect, are hydroxide ion (pK_a 15.7) compared with hydroperoxide ion (pK_a 11.6), and ammonia (pK_a 9.3) compared with hydrazine (pK_a 8.0). In each of these pairs the weaker base is the stronger nucleophile.

Solvation of nucleophilic anions markedly influences their reactivity. The nucleophilicities cited above were obtained from reactions in methanol solution. Polar, protic solvents such as water and alcohols solvate anions by hydrogen bonding interactions, as shown in the diagram on the right. These solvated species are more stable and less reactive than the unsolvated "naked" anions. Polar, aprotic solvents such as DMSO (dimethyl sulfoxide), DMF (dimethylformamide) and acetonitrile do not solvate anions nearly as well as methanol, but provide good solvation of the accompanying cations. Consequently, most of the nucleophiles discussed here react more rapidly in solutions prepared from these solvents. Such solvent effects are more pronounced for small basic anions than for large weakly basic anions. Thus, for reaction in DMSO solution we observe the following reactivity order:

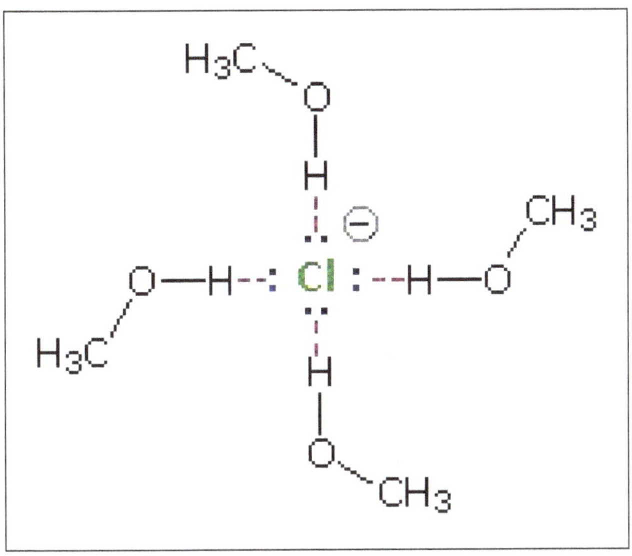

Nucleophilicity: $I^{(-)} < Br^{(-)} < Cl^{(-)} \sim N_3^{(-)} < CH_3CO_2^{(-)} < CN^{(-)} \sim CH_3S^{(-)} < CH_3O^{(-)}$

Unlike the order in methanol, this ranking is roughly the order of increasing basicity. Neutral bases like ammonia and simple primary and secondary amines have similar basicities (pKa = 9 to 11) and are also good nucleophiles. In reactions with methyl bromide in methanol their nucleophilicity is greater than $Cl^{(-)}$ and less than $N_3^{(-)}$. In slightly basic aprotic solvents they react with moderate electrophiles such as alkyl halides (Cl, Br and I) and carbonyl groups (aldehydes, ketones, esters and anhydrides), giving quaternary salts from the former, and imines, enamines and amides from the latter. Tertiary amines and bulky secondary amines are poorer nucleophiles and behave exclusively as bases.

Solvent Effects

Acids

The ionization of a group of acids in a given solvent may be used as a measure of acid strength, and in such cases the solvent serves as a common reference base. The determination of K_a and pK_a values for any solvent system may be carried out in the same way as in water; however, the acidities obtained for a group of acids measured in different solvents will generally be different, both in the numerical value for each, and sometimes in relative order.

$$\underset{\text{an acid}}{A-H} + \underset{\text{solvent}}{Sol:} \rightleftarrows \underset{\text{conjugate base}}{A:^{(-)}} + \underset{\text{conjugate acid}}{Sol-H^{(+)}}$$

There are several factors which influence acidity measurements made in water and other solvents. These are:

- Ionization of an acid produces ions. Solvents having large dielectric constants favor charge formation and separation.
- The conjugate acid from the solvent is a cation that is stabilized by association with other solvent molecules, a phenomenon called solvation.
- The conjugate base of the acid is an anion that in most cases is also stabilized by solvation.

As examples, consider the following data obtained for six strong acids in three common solvents.

Table: Experimental pK_a values for some strong acids in different solvents.

Compound	H_2O	DMSO	CH_3CN
$HClO_4$	−10*	−10*	1.6
CF_3SO_3H	−10*	−10*	2.6
HBr	−9	0.8	5.8
HCl	−7	1.8	10.4
H_2SO_4	−3	1.4	7.9
CF_3CO_2H	−0.25	3.5	12.7

*Perchloric acid and trifluoromethanesulfonic acid are both completely ionized in water and DMSO.The pK_a value of −10 is approximate, and is intended to reflect the high degree of dissociation.

All three solvents have fairly high dielectric constants, but water ($\varepsilon = 80$) is significantly greater than DMSO ($\varepsilon = 46.7$) or acetonitrile ($\varepsilon = 37.5$). The greater acidity (lower pK_a)

of the last four compounds in water might be attributed to the dielectric difference; however, the superior anion solvation provided by water is considered to be the major factor. Both DMSO and acetonitrile are poor anion solvation solvents, consequently the ionization equilibrium will be shifted to the left.

It is informative to focus first on the pK_a differences between water and DMSO. We do this because water and DMSO have similar basicities, so stabilization of the solvent conjugate acid species should not differ substantially. HBr and HCl display pK_as in DMSO that are roughly 9 to 10 powers of ten more positive than in water, and the weaker second period acid HF shows an even greater change (pK_a = 3.17 in water & 15 in DMSO). The conjugate bases of these acids are all single atom anions. Because the negative charge is localized, anion stabilization by solvation is probably the most important factor in establishing the position of these equilibria. In contrast, the conjugate bases from sulfuric acid and trifluoroacetic acid are internally stabilized by charge delocalization over two or three oxygen atoms. The importance of solvation is therefore less important, and the shift in acidity in DMSO is reduced to 4 powers of ten.

When acetonitrile is the solvent, the measured pK_as of these acids show a further increase, and even the strongest acids are not completely dissociated. This change may be attributed in large part to diminished solvation of the solvent conjugate acid, which is the chief cation species in dilute solutions. Since both the cationic and anionic species from acidic ionization are poorly stabilized by solvation, the dissociation equilibrium lies far to the left (little ionization).

Table: Experimental pK_as for some weak acids in different solvents.

Compound	H_2O	DMSO	CH_3CN
C_2H_5OH	17	29	42
CF_3CH_2OH	12.4	23.5	33.5
$C_6H_5C{\equiv}CH$	19	29	40.6
C_4H_9SH	10.5	17	28.6
CH_3CO_2H	4.8	12.8	22.3
$C_6H_5CO_2H$	4.2	11.1	20.7
C_6H_5OH	10	18	26.6
C_6H_5SH	6.6	10.3	20.6

Most organic compounds are much weaker acids than the strong acids. It is therefore of interest to determine the changes in acidity that take place when such compounds are examined in the same three solvents. The first four compounds in this table (shaded light green) are weak acids having conjugate bases in which the charge is localized on a single atom (O, C & S). Changing the solvent from water to DMSO decreases the acidity by 10 to 12 powers of ten for second period atoms (C & O), and by 6.5 for the third period sulfhydryl acid. Just as in the more acidic cases treated above, solvation of the conjugate base anions is an important stabilizing factor in water, although the lower charge density on sulfur reduces its significance. The further increase in pK_a that occurs in acetonitrile solutions, relative to DMSO, is fairly uniform (10 to12 units) and a bit larger than the shift found for the strong acids.

The last four compounds (shaded blue) have conjugate bases stabilized by charge delocalization. Here, the decrease in acidity going from water to DMSO is lowered to 7 to 8 powers of ten for oxygen and is less than 4 for sulfur. As noted previously, charge delocalization within an anion reduces the importance of solvation stabilization. The increase in pK_a found for acetonitrile solutions remains roughly the same, and may be attributed to poor solvation in this solvent.

With the exception of the acetylene derivative, the previous compounds are all heteroatom acids. Similar measurements for a group of activated carbon acids show agreement with the previous analysis, and illustrate the manner in which anionic charge delocalization reduces the pK_a difference between water and DMSO measurements. Note the near identity of the second and third examples in the following table. As expected, the pK_a increase in going to acetonitrile from DMSO remains roughly constant.

Table: Experimental pK_as for some carbon acids in different solvents.

Compound	H_2O	DMSO	CH_3CN
$(CH_3CO)_2CH_2$	9	13.3	24.4
$(N{\equiv}C)_2CH_2$	11.2	11.2	22.2
$(C_6H_5SO_2)_2CH_2$	11.6	12.2	23.2
$(C_2H_5OCO)_2CH_2$	13.5	16.4	27.8
$C_4H_4CH_2$ cyclopentadiene	15.0	18.0	31.2

Bases

The most common notation for reporting relative base strengths of a neutral base (B:) is in terms of the pK_a's of the corresponding conjugate acids (these conjugate acids are often called "onium" cations). The following equation shows the equilibrium involved in this relationship. Note that strong bases will have weakly acidic conjugate acids, so the pK_a is proportional to the base strength of the neutral base.

$$\underset{\text{conjugate acid}}{B-H^{(+)}} + \underset{\text{solvent}}{\text{Sol}} \rightleftharpoons \underset{\text{a base}}{B:} + \underset{\text{conjugate acid}}{\text{Sol}-H^{(+)}}$$

The previously defined factors that influenced acidity may now be reexamined for this new equilibrium.

- Ions are present on both sides of the equation. Consequently, differences in the solvent dielectric constant should be a relatively unimportant factor.

- The base and solvent conjugate acid cations are both stabilized by solvation. This will be a critical factor, and is related to solvent basicity.
- There are no anions in the above equation; however charge neutralization requires a counter anion. Since the same anion will be present on both sides of the equation, its influence on this equilibrium will be cancelled.

The following table gives pK_a values for some important nitrogen bases, some structures for which are shown to its right. In the triethylamine and DABCO examples at the top of the table, the cationic charge is relatively localized on a single nitrogen atom. The charge in lutidine may be delocalized onto ring carbons at the cost of aromatic stabilization. The last three examples are compounds in which the charge is delocalized by resonance (protonation occurs at the light blue nitrogen) or internal hydrogen bonding (the proton sponge).

In contrast to the earlier examples of acid pK_as, the values for these ammonium cations are nearly identical in water and DMSO solvents. Indeed, the fact that most of the DMSO pK_as are a bit lower than their water counterparts suggests that DMSO is slightly more basic than water. In acetonitrile all the pK_a values are about 10 units higher than the DMSO values.

Table: pK_a s for some ammonium species in different solvents.

Cation	H_2O	DMSO	CH_3CN
$(C_2H_5)_3N{-}H^{(+)}$	10.75	9.0	18.5
DABCO–$H^{(+)}$	8.8	8.9	18.3
Lutidine–$H^{(+)}$	6.3	4.45	14
DBU–$H^{(+)}$	12	12	24.3
$[(CH_3)_2N]_2C{=}NR{-}H^{(+)}$	13.8	13.6	23.3
Proton Sponge–$H^{(+)}$	12.3	7.5	ca. 18

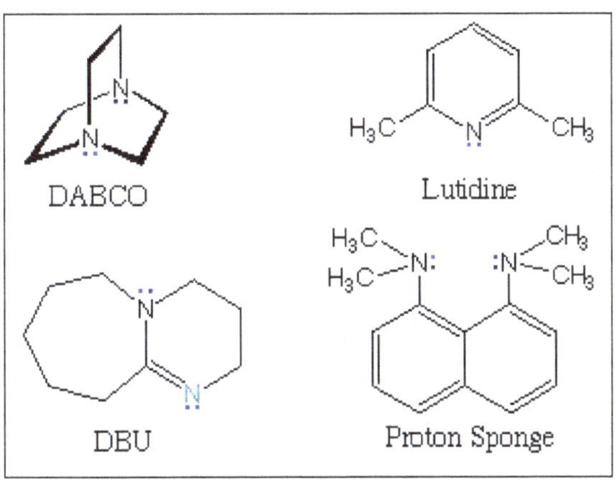

The influence of solvent on acidity and basicity noted above may cause unexpected changes in simple acid-base equilibria. If acetic acid and triethyl amine are mixed together in water, a rapid proton transfer takes place to give nearly quantitative formation of triethyl ammonium acetate. The conjugate acids differ in strength by 10^6 and the weakest is favored at equilibrium. When the same two compounds are combined in DMSO solution there is negligible proton transfer, since acetic acid is now over 10^3 times weaker an acid than the triethyl ammonium cation.

Hammett Acidity Function

Ionization measurements of aqueous solutions of acids (and bases) are generally limited to a pH range of 1.5 to 13. Relative values of stronger acids (and bases) require the use of other solvents, and due to variations in dielectric constants, solvation, and other deviations from ideality, consistant relationships are not achieved. A remedy for this situation was proposed by Louis Hammett (Columbia University). Hammett suggested using a series of weak indicator bases wherein the $[B]/[BH^+]$ ratio could be determined colorimetrically due to a distinctively colored conjugate acid. When the ratio of [B] to $[BH^+]$ becomes especially small and experimentally impossible to measure accurately, a different, less basic indicator, overlapping the first, would serve to extend the acidity range. Additional weaker indicator bases would then allow one to leapfrog across the acidity scale, extending it roughly 10^{10} fold. To this end, an acidity function, H_0, was defined as shown by the following equation:

$$H_0 = pK_{BH^+} + \log \frac{[B]}{[BH^+]}$$

Idealy, H_0 provides a quantitative measure of acid strength for non-aqueous and concentrated aqueous Brønsted acids. An acidity function is a measure of the acidity of a medium or solvent system, usually expressed in terms of its ability to donate protons to a base of known basicity. The predominant acid species in dilute aqueous solutions is H_3O^+, so the pH scale is by far the most commonly used acidity function in such systems. In concentrated sulfuric acid the predominant acid species is $H_3SO_4^+$, a much stronger acid, thus H_0 is an extension of the pKa scale for very strong acids. A key factor for measuring a meaningful H_0 is the choice of a suitable solvent. Ideally. a weakly basic, protic solvent, having a large dielectric constant, a significant degree of autoprotolysis and good solvation of ions is desired. Hammett's unconventional choice was anhydrous sulfuric acid. Despite obvious concerns of corrosiveness, hygroscopicity and viscosity, this proved to be an inspired selection, as the following characteristics demonstrate. Since traces of water are undesirable, titration with SO_3 (fuming sulfuric acid) to a melting point maximum is necessary.

Properties of Anhydrous Sulfuric Acid				
Autoprotolysis	Melting point	Autoprotolysis constant	Dielectric constant	pKa

$2H_2SO_4 \rightleftarrows H_3SO_4^{(+)} + HSO_4^{(-)}$	10.3 °C	$K_{ap} = 2.7\ 10^{-4}$	~100	−2.8

The following tables list some of Hammett's indicator bases and the H_o's of some very strong acids or "super acids". The pKa of the indicator bases were determined in aqueous solutions of the strong acids H_2SO_4 and $HClO_4$. Ho values were determined in anhydrous sulfuric acid, and those stronger than -16 are estimates.

Hammett Indicator Constants (pKa)		
Indicator	H_2SO_4/Water	$HClO_4$/Water
p-Nitroanilin e	+1.11	+1.11
o-Nitroaniline	-0.20	-0.19
p-Nitrodiphenylaniline	-2.38	---
2,4-Dichloro-6-nitroaniine	-3.22	---
2,4-Dinitroaniline	-4.38	-4.33
Benzalacetophenone	-5.61	---
p-Benzoyldiphenyl	-6.19	---
6-Bromo-2,4-dinitroaniline	-6.59	---
Anthraquinone	-8.15	---
2,4,6-trinitroaniline	-9.29	---

Ho for Some Strong Acids (pKa)	
Acid	Ho
Sulfuric Acid	-12.0
Perchloric Acid	-13.0
Chlorosulfonic Acid	-13.8
Triflic Acid (CF_3SO_3H)	-14.6
Hydrogen Fluoride	-15.1
Fluorosulfonic Acid	-15.6
Carborane Superacids $H(CXB_{11}Y_5Z_6)$	~ -19
Magic Acid ($FSO_3H{\cdot}SbF_5$)	-21
Fluoroantimonic Acid ($HSbF_6$)	-21

Strong Bases

Organic chemists need bases of varying strength as reagents tailored to the requirements of specific reactions. Neutral nitrogen compounds can cover a basicity range of roughly 10^{10}, as shown in the following table. Pyridine is commonly used as an acid scavenger in reactions that produce mineral acid co-products. Its basicity and nucleophilicity may be modified by steric hindrance, as in the case of lutidine ($pK_a = 6.3$), or resonance stabilization, as in the case of 4-dimethylaminopyridine ($pK_a = 9.2$). Hünig's base is relatively non-nucleophilic (due to steric hindrance), and like DBU is often used as the base in E2 elimination reactions conducted in non-polar solvents. Barton's base is a strong, poorly-nucleophilic, neutral base that serves in cases where electrophilic substitution of DBU or other amine bases is a problem.

Base Name	Pyridine	Lutidine	DABCO	DMAP	Triethyl Amine	Hünig's Base	DBU	Proton Sponge	Barton's Base
Formula	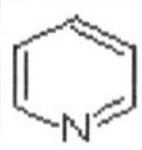	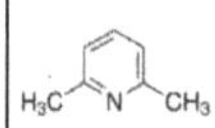		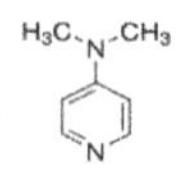	$(C_2H_5)_3N$	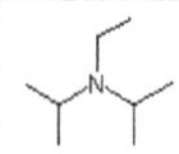	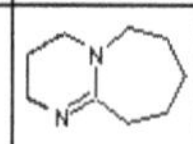	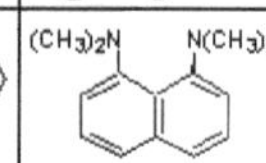	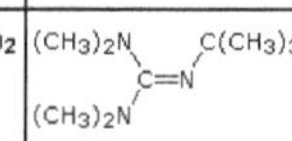
pK_a (onium)	5.3	6.27	8.8	9.2	10.7	11.4	12	12.4	14

The strongest bases available to organic chemists are alkali metal alkoxide salts such as potassium tert-butoxide (pK_a = 16 to 19), alkali metal alkyls such as Grignard reagents, n-butyl lithium and dimsyl sodium (pK_a = 28 to upper 40s), and amide salts of alkali metals, such as LDA (pK_a = 30 to 40) and sodium hydride (pK_a ~ 42). These powerful bases are all potential nucleophiles (some more than others) and have partially ionic bonds to the metals. Bases in these pKa ranges are used to convert weak C-H acids into useable carbanionic-metal reagents.

Weak C-H Acids	
Compound	pK_a
Nitromethane	10.2
Diethylmalonate	13
Cyclopentadiene	16
Acetone	20
Acetylene	25
Acetonitrile	25
Dimethylsulfone	28
DMSO*	35
Triphenylmethane	33
Toluene	41
Benzene	43
Ethane	49

* DMSO = Dimethylsulfoxide

Some examples of alkylmetal preparations, together with similar amide metal bases are shown in the following equations. In all cases inert solvents such as benzene, cyclohexane, tetrahydrofuran and ether are used. Since these reactions proceed to completion, equilibrium arrows are not drawn.

- Methyl Magnesium Iodide: $CH_3I + Mg$ —> CH3MgI (ether solution).
- Butyl Lithium: $2\ Li + C_4H_9X$ —> $C_4H_9Li + LiX$ (where X = Cl, Br).
- Trityl Sodium*: $(C_6H_5)_3CCl + 2\ Na$ —> $(C_6H_5)_3CNa + NaCl$.
- Dimsyl Sodium: $CH_3SOCH_3 + NaH$ —> $CH_3SOCH_2Na + H_{2,}$
- LDA: $[(CH_3)_2)CH]_2NH + C_4H_9Li$ —> $[(CH_3)_2)CH]_2NLi + C_4H_{10.}$
- LiHMDS: $\{(CH_3)_3Si\}_2NH + C_4H_9Li$ —> $\{(CH_3)_3Si\}_2NLi + C_4H_{10.}$

*This anion is red and serves as an indicator. It is also a one electron reductant.

Nitrile Alkylation: $RCH_2CN + LDA$ —> $RCHCNLi + \{(CH_3)_2)C\}_2NH$;

RCHCNLi + R'X —> R'RCHCN + LiX

Aldehydes, ketones and esters having activated C-H groups may be similarly converted to their conjugate bases, followed by alkylation or addition to electrophilic functions.

Weak N-H Acids	
Compound	pK_a
Pyrrole	16
Aniline	27
HMDS*	27
Ammonia	34
Pyrrolidine	35
Diisopropyl Amine	35.7
TMP*	37
Hydrogen, H_2	~ 42

*TMP = 2, 2, 6, 6-Tetramethylpiperidine

*HMDS = Bis(trimethylsilyl)amine

Nonionic Superbases

Recently, a new class of non-metallic, poorly nucleophilic, neutral bases have been prepared and studied. Some examples are shown in the following diagram.

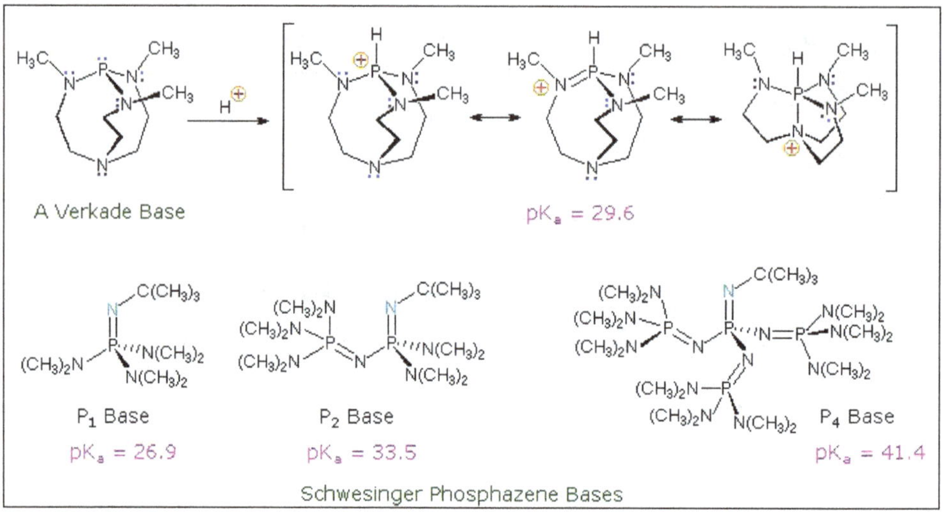

The basic site in the Verkade base is the phosphorous atom, the conjugate acid being stabilized by transannular bonding to nitrogen. The strength of these bases may be modified by substituents on the flanking nitrogens. The Schwesinger phosphazene bases increase their strength as additional phosphazene units are added in conjugation with the basic site (the light blue nitrogen atom). All the pK$_a$s for these bases are measured in acetonitrile.

Kinetic Acidity

The most common acid-base terminology, pK$_a$, reflects an equilibrium acidity, extrapolated or normalized to water. In the following equation a base, B:$^{(-)}$ M$^{(+)}$, abstracts a proton from an acid, H-A, to form a conjugate acid - base pair (A:$^{(-)}$ M$^{(+)}$ & B-H). The rate of the forward proton abstraction is k_f, and the reverse rate of proton transfer is k_r. This kind of equilibrium is usually characterized by an equilibrium constant, K_{eq}, which is the ratio of the rate constants (k_f / k_r). If H-A is a weaker acid than H-B the equilibrium will lie to the left, and K_{eq} will be smaller than 1.

$$\underset{(\text{acid}^1)}{\text{H}-\text{A}} + \underset{(\text{base}^1)}{\text{B}:^{(-)}\text{M}^{(+)}} \underset{k_r}{\overset{k_f}{\rightleftharpoons}} \underset{(\text{base}^2)}{\text{A}:^{(-)}\text{M}^{(+)}} + \underset{(\text{acid}^2)}{\text{B}-\text{H}}$$

In cases where H-A is very much weaker than H-B, K_{eq} may be too small to measure, but it may be possible to determine the rate of the forward proton abstraction under certain circumstances. If an isotopically labeled conjugate acid of the base is used as a solvent for the reaction (B-D in the following equations), then any proton abstraction that occurs will be marked by conversion of H-A to D-A. The green shaded top equation shows the initial loss of the proton, and the second equation describes the rapid deuteration of the intermediate conjugate base, $A:^{(-)}$. As these reactions proceed, the H-A reactant will be increasingly labeled as D-A, and the rate of isotope exchange will indicate the kinetic acidity of H-A. It is assumed that kinetic acidity is roughly proportional to equilibrium (thermodynamic) acidity, but this is not always true.

$$\text{solvent} = B-D$$

$$H-A+B:^{(-)}M^{(+)} \rightleftarrows A:^{(-)}M^{(+)}+B-H$$

$$D-A+B:^{(-)}M^{(+)} \rightleftarrows A:^{(-)}M^{(+)}+B-D$$

The following diagram provides an instructive example of these principles. The first equation, in the yellow shaded box, provides important information about heavy water (deuterium oxide). Heavy water is similar to water in many respects, but is 10% more dense and a ten-fold weaker acid. A 1 molar concentration of sodium deuteroxide will serve as the base, and an equimolar quantity of 3, 3-dimethyl-1-butyne will serve as the weak acid. The most acidic hydrogen in this hydrocarbon (colored red) is at C-1. In practice, we would need to use a co-solvent to completely dissolve the hydrocarbon in the heavy water, but this has been omitted in order to simplify the discussion.

The second equation describes the essential changes expected on combining these reactants in the heavy water solvent. Since the terminal alkyne is a much weaker acid than heavy water, acid-base equilibria do not favor its conjugate base. Nevertheless, if the acetylide anion is formed, even in low concentration, it should react quickly by abstracting a deuterium from a neighboring deuterium oxide molecule. The result would be an observable exchange of deuterium for hydrogen, testifying that an acid-base reaction has occurred.

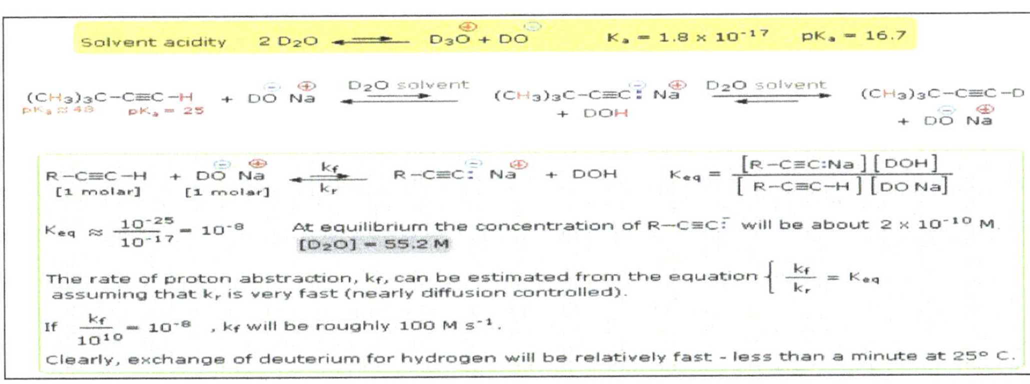

$K_{eq} \approx \dfrac{10^{-25}}{10^{-17}} = 10^{-8}$ At equilibrium the concentration of R–C ≡ : will be about 2 × 10^{-10} M.

The rate of proton abstraction, k_f, can be estimated from the equation $\left\{ \dfrac{k_f}{k_r} = K_{eq} \right.$ assuming that k_r is very fast (nearly diffusion controlled).

If $\dfrac{k_f}{10^{10}} = 10^{-8}$, k_f will be roughly 100 Ms^{-1}.

Clearly, exchange of deuterium for hydrogen will be relatively fast – less than a minute at 25 °C.

The green shaded box contains equations that help us to interpret the experimental results. In order to evaluate the equilibrium acidity of the substrate, we would need to measure the equilibrium constant K_{eq} for the initial acid-base equilibrium, shown at the top of the shaded box. Since we know the K_a 's of 3,3-dimethyl-1-butyne and heavy water, we can estimate K_{eq} by dividing the former (10^{-25}) by the latter (10^{-17}). This calculation reveals a K_{eq} that would be difficult to measure directly because of its small magnitude (10^{-8}). Indeed, the equilibrium concentration of acetylide anion is estimated to be only $2*10^{-10}$ M.

If we examine this experiment from the viewpoint of kinetics, easily observable evidence of terminal alkyne acidity is obtained. The last three rows of equations in the green shaded box make this clear. Since K_{eq} is the ratio of forward and reverse rate constants, it is possible to draw conclusions about the rate of terminal proton abstraction from the alkyne. This leads to the conclusion that reasonably rapid hydrogen-deuterium exchange will occur, even though the acetylide anion is never present in concentrations exceeding 10^{-9} M.

This example also demonstrates the limits of the isotope exchange approach. The 3, 3-dimethyl-1-butyne substrate also has nine other hydrogen atoms (colored orange) that do not exchange with deuterium under these conditions. We know that these hydrogens are much less acidic (K_a ca. 10^{-48}), and it is interesting to consider their potential participation in acid-base reactions by the previous analysis. The estimated K_{eq} for such carbanion formation is ca.10^{-30}, taking into account the nine-fold increase in concentration. This implies a concentration of one carbanion in every 10^9 liters of solution. The kinetic analysis is equally discouraging. The forward rate constant is estimated to be 10^{-20} Ms^{-1}. The time required to exchange half these hydrogens for deuterium would therefore be about 100 centuries.

In order to study the kinetic acidity of extremely weak acids (pK_a's = 30 to 50) it is necessary to use much stronger bases, which of course have much weaker conjugate acids. Amide anions (pK_a's = 26 to 36) have been used for this purpose.

By comparing the rates of hydrogen exchange for different compounds under identical conditions, tables of relative kinetic acidities may be assembled. An interesting example of such a study has been reported for a group of nitroalkanes having acidic α-hydrogens. Removal of an α-hydrogen by a base generates a conjugate base called an aci-anion, as shown here.

$$\underset{\text{nitro compound}}{R_2CH-NO_2} + B:^{(-)}M^{(+)} \rightleftarrows \underset{\text{aci-conjugate base}}{R_2C=NO_2^{(-)}M^{(+)}} + B-H$$

Compound	pK_a	Relative Rate of Exchange
CH_3NO_2	10.2	120
$CH_3CH_2NO_2$	8.5	20
$(CH_3)_2CHNO_2$	7.7	1.0

Compared with the terminal alkyne, such nitroalkanes are relatively strong C-H acids. Since they are stronger acids than water, the kinetic exchange experiments must be conducted under milder conditions than those used for the terminal alkyne. This is achieved by using smaller base concentrations and lowering the temperature of the exchange reaction. Accurate pK_a's of 2-nitropropane, nitroethane and nitromethane may be measured directly in aqueous solution. These kinetic and equilibrium acidities are listed in the table on the right. Note that for these three compounds, kinetic acidity changes in an opposite fashion to equilibrium acidity. The kinetic order seems to reflect steric hindrance and carbanion stability; whereas, the equilibria favor increased substitution of the aci-anion double bond.

Base-catalyzed isotope exchange studies of compounds incorporating more than one set of acidic hydrogens provides additional insight concerning the creation and use of nucleophilic conjugate bases. Ketones provide many examples of regioisomeric enolate base formation, and the following diagram shows two such cases. As noted in the nitroalkane study, hydrogens on an α-methyl group are exchanged more rapidly than those on more substituted α-carbon atoms. The equations in the diagram show only the initial product from a single exchange. These products have additional α-hydrogens which are also exchanged by subsequent reactions of this kind, so that complete replacement of all α-hydrogens by deuterium takes place in a short time.

The relative stability of the resulting enolates increases with substitution of the enolate double bond. In order to determine enolate anion equilibria for these ketones, the bulky strong base sodium Bis(trimethylsilyl)amide (pK_a = 26) was used.

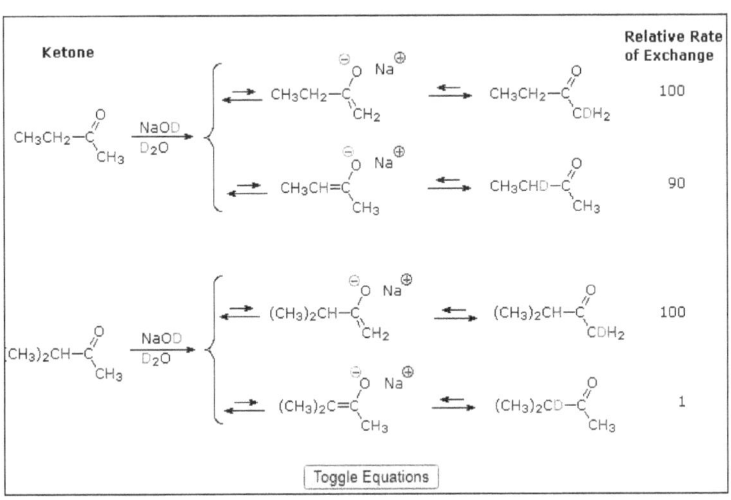

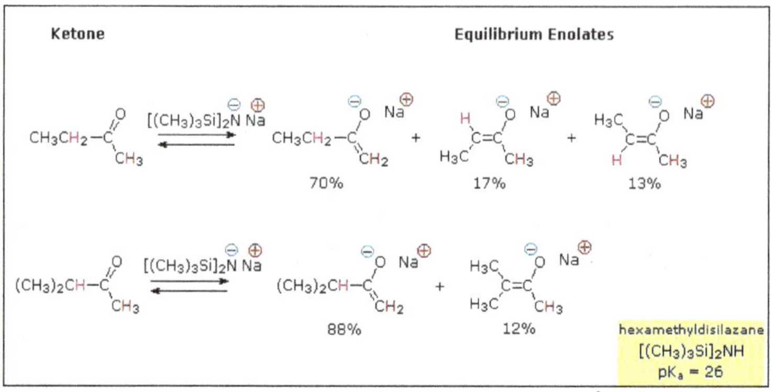

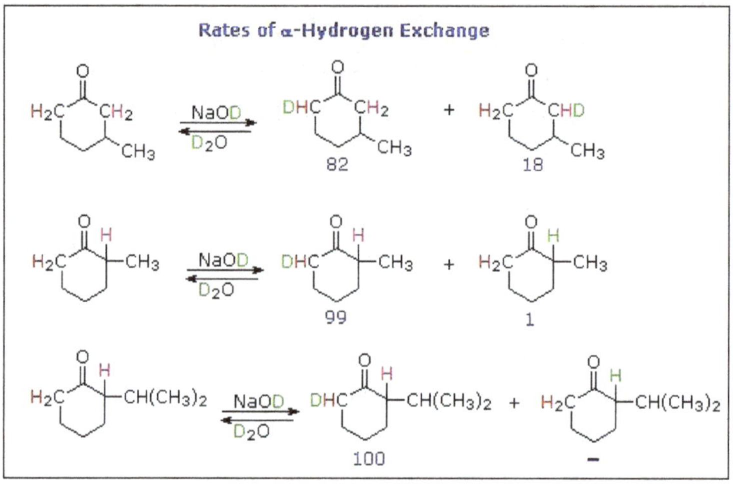

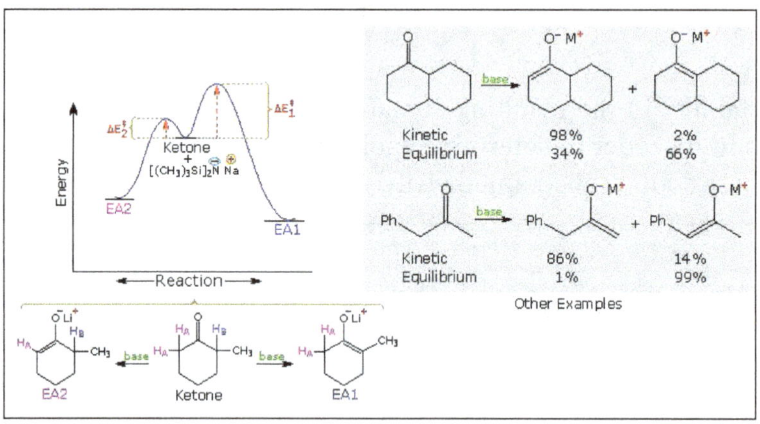

Most carbon acids yield conjugate bases that are stabilized by charge delocalization onto neighboring heteroatoms. This resonance stabilization requires significant structural reorganization of the initial compound, which in turn imposes an energy barrier that retards the rate of proton abstraction. For example, the alpha-carbon of a ketone or ester must undergo rehybridization as the enolate anion is formed. The stereoelectronic demands of this change were described above, and it is not surprising that enolate anion formation is much slower than equivalent proton transfers between alcohols and other hydroxylic compounds. Deprotonation rates of phenol and nitromethane, compounds with nearly identical pK_a's (10.0), provide an instructive example of this structural reorganization factor. The acidic proton in phenol is bound to oxygen, so deprotonation requires little structure change and is very fast. Nitromethane is a carbon acid. Deprotonation to an aci-anion involves considerable structural change, and is a million times slower than phenolate formation. These structural changes are illustrated in the following diagram.

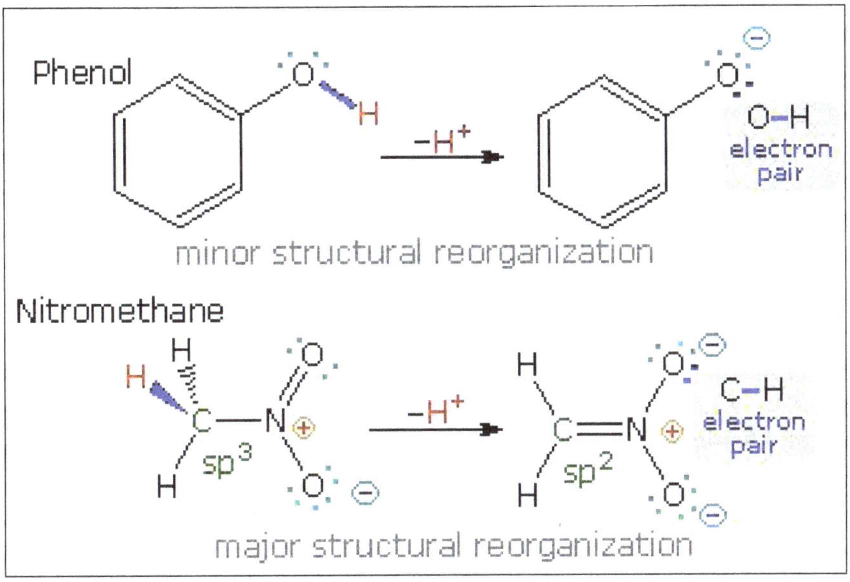

Note that the O-H electron pair in phenol remains largely on oxygen in the corresponding conjugate base, whereas the C-H electron pair in nitromethane is predominantly shifted to oxygen in its conjugate base (colored blue). The trends outlined here are a bit oversimplified, since solvent and cation influences have been ignored.

Equilibrium

Chemical equilibrium is the state in which the forward reaction rate and the reverse reaction rate are equal. In theory, all chemical reactions are in fact double reactions: for every forward reaction, there is a subsequent reverse reaction. The idea can be illustrated as follows:

reactants $\rightleftharpoons$ products

For plenty of reactions, however, the forward reaction is so favored, and the reverse reaction is so negligible, that reactions are written simply in terms of the solid forward arrow, $A \rightarrow B$. However, we will now consider forward/reverse reaction pairs that exist in chemical equilibrium with one another.

Chemical Equilibrium

Cars entering and leaving a city over a bridge: Chemical equilibrium is akin to two cities connected a bridge. If the rate at which cars leave the city is exactly equal to the rate at which cars enter the city, the result is that the net number of cars on both sides of the bridge does not change.

In a chemical reaction, chemical equilibrium is the state in which the forward reaction rate and the reverse reaction rate are equal. The result of this equilibrium is that the concentrations of the reactants and the products do not change. However, just because concentrations aren't changing does not mean that all chemical reaction has ceased. Just the opposite is true; chemical equilibrium is a dynamic state in which reactants are being converted into products at all times, but at the exact rate that products are being converted back into reactants. The result of such a situation is analogous to a bridge between two cities, where the rate of cars going over the bridge in each direction is exactly equal. The result is that the net number of cars on either side of the bridge does not change.

Reaching Equilibrium and the Equilibrium Position

When a system consists of competing forward and reverse reaction rates, the reaction will proceed until chemical equilibrium is reached. Consider the following general reaction between A and B:

$$A \rightleftharpoons B$$

Let's say that A and B are introduced into a reaction vessel, and a reaction begins to proceed. The concentrations of A and B will change over time, until they reach equilibrium.

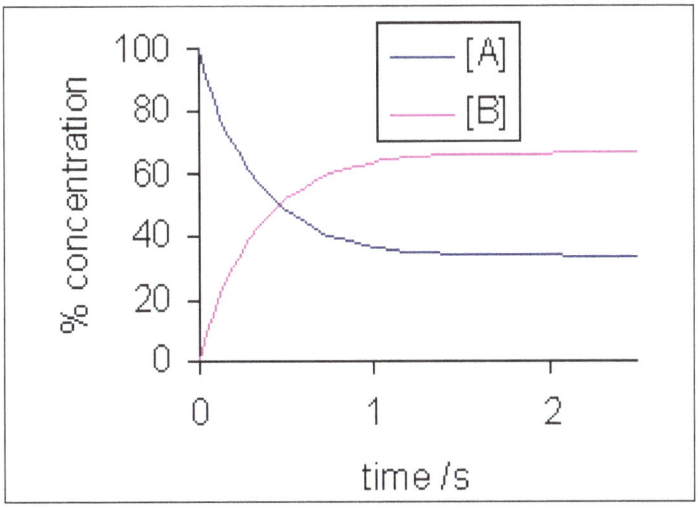

Concentration vs. time for a reversible reaction: Note that equilibrium is reached when both the curves plateau, and the concentrations of both reactants and products do not change afterward.

Note that over time, the curves level out, or plateau, and the concentrations of A and B are no longer changing. This is the point at which the system has reached chemical equilibrium. While there are various factors that can increase or decrease the amount of time it takes for a given system to reach equilibrium, the equilibrium position itself is unaffected by these factors. For instance, if a catalyst is added to the system, the reaction will proceed more quickly, and equilibrium will be reached faster, but the concentrations of both A and B will be the same at equilibrium for both the catalyzed and the uncatalyzed reaction.

Chemical Laws

Navigating the world of chemistry is much easier once you have an understanding of the field's basic laws. Here are brief summaries of the most important laws, the foundational concepts, and principles of chemistry:

Avogadro's Law

Equal volumes of gases under identical temperature and pressure will contain equal numbers of particles (atoms, ions, molecules, electrons, etc.)

Boyle's Law

At a constant temperature, the volume of a confined gas is inversely proportional to the pressure to which the gas is subjected:

$$PV = k$$

Charles' Law

At a constant pressure, the volume of a confined gas is directly proportional to the absolute temperature in Kelvin:

$$V = kT$$

Conservation of Energy

Energy can neither be created nor destroyed; the energy of the universe is constant. This is the First Law of Thermodynamics.

Conservation of Mass

Matter can neither be created nor destroyed, though it can be rearranged. Mass remains constant in an ordinary chemical change. This principle is also known as Conservation of Matter.

Dalton's Law

The pressure of a mixture of gases is equal to the sum of the partial pressures of the component gases.

- Definite Composition: A compound is composed of two or more elements chemically combined in a defined ratio by weight.
- Dulong-Petit Law: Most metals require 6.2 calories of heat to raise the temperature of one gram-atomic mass of metal by one degree Celsius.

Faraday's Law

The weight of any element liberated during electrolysis is proportional to the quantity of electricity passing through the cell and also to the equivalent weight of the element.

First Law of Thermodynamics

The total energy of the universe is constant and can neither be created nor destroyed. This law is also known as Conservation of Energy.

- Gay-Lussac's Law: The ratio between the combining volumes of gases and the product (if gaseous) can be expressed in small whole numbers.

Graham's Law

The rate of diffusion or effusion of a gas is inversely proportional to the square root of its molecular mass.

Henry's Law

The solubility of a gas (unless it is highly soluble) is directly proportional to the pressure applied to the gas.

Ideal Gas Law

The state of an ideal gas is determined by its pressure, volume, and temperature according to the equation:

$$PV = nRT$$

where P is the absolute pressure, V is the volume of the vessel, n is the number of moles of gas, R is the ideal gas constant, and T is the absolute temperature in Kelvin.

- Multiple Proportions: When elements combine, they do so in the ratio of small whole numbers. The mass of one element combines with the fixed mass of another element according to certain ratios.
- Periodic Law: The chemical properties of the elements vary periodically according to their atomic numbers.
- Second Law of Thermodynamics: Entropy increases over time. Another way of stating this law is to say that heat cannot flow, on its own, from an area of cold to an area of hot.

References

- Chemistry: britannica.com, Retrieved 08, March 2020
- Elements-and-compounds: lumenlearning.com, Retrieved 13, August 2020
- Phases-and-Their-Transitions, Physical-Equilibria, Physical-and-Theoretical-Chemistry-Textbook-Maps, Supplemental-Modules-(Physical-and-Theoretical-Chemistry): libretexts.org, Retrieved 27, January 2020
- Definition-of-bonds-in-chemistry-604392: thoughtco.com, Retrieved 05, April 2020
- Types-of-energy: lumenlearning.com, Retrieved 09, July 2020
- Chemical-reactions: byjus.com, Retrieved 20, February 2020

2

Theories of Chemistry

There are a number of theories which are studied within chemistry such as atomic theory, molecular orbital theory, transition state theory, Marcus-Hush theory, valence bond theory and VSEPR theory. This chapter has been carefully written to provide an easy understanding of the basic concepts associated with these theories of chemistry.

Atomic Theory

John Dalton is the scientist credited for proposing the atomic theory. This theory explains several concepts that are relevant in the observable world: the composition of a pure gold necklace, what makes the pure gold necklace different than a pure silver necklace, and what occurs when pure gold is mixed with pure copper. Before discussing the atomic theory, here we explain the theories that Dalton used as a basis for his theory: the law of conservation of mass and the law of constant composition.

Law of Conservation of Mass

The law of conservation of mass states that the total mass present before a chemical reaction is the same as the total mass present after the chemical reaction; in other words, mass is conserved. The law of conservation of mass was formulated by Antoine Lavoisier as a result of his combustion experiment, in which he observed that the mass of his original substance—a glass vessel, tin, and air—was equal to the mass of the produced substance—the glass vessel, "tin calx", and the remaining air.

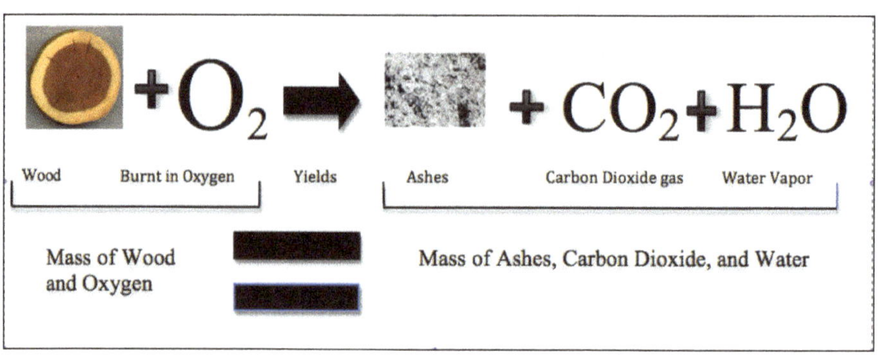

Historically, this was a difficult concept for scientists to grasp. If this law was true, then how could a large piece of wood be reduced to a small pile of ashes? The wood clearly has a greater mass than the ashes. From this observation scientists concluded that mass had been lost. However, the illustration below shows that the burning of word does follow the law of conservation of mass. Scientists did not take into account the gases that play a critical role in this reaction.

Law of Constant Composition

Joseph Proust formulated the law of constant composition (also called the law of definite proportions). This law states that if a compound is broken down into its constituent elements, the masses of the constituents will always have the same proportions, regardless of the quantity or source of the original substance. Joseph Proust based this law primarily on his experiments with basic copper carbonate. The illustration below depicts this law; 31 grams of H_2O and 8 grams of H_2O are made up of the same percent of hydrogen and oxygen.

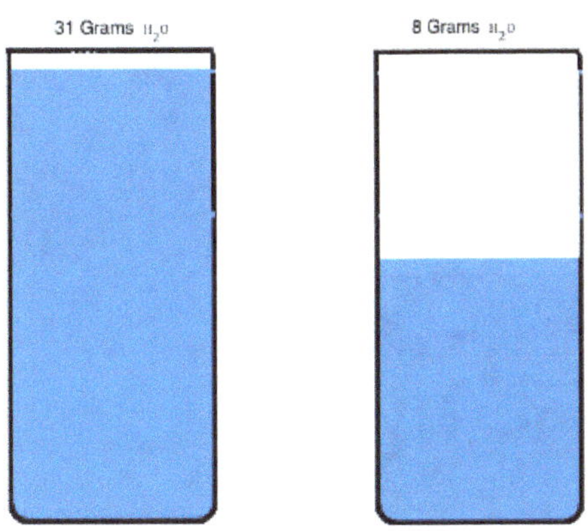

Compound	Grams of H	Grams of O	Percent H	Percent O
31 grams H_2O	3.469	27.53	11.19%	88.81%
8 grams H_2O	.8951	7.105	11.19%	88.81%

Dalton's Atomic Theory

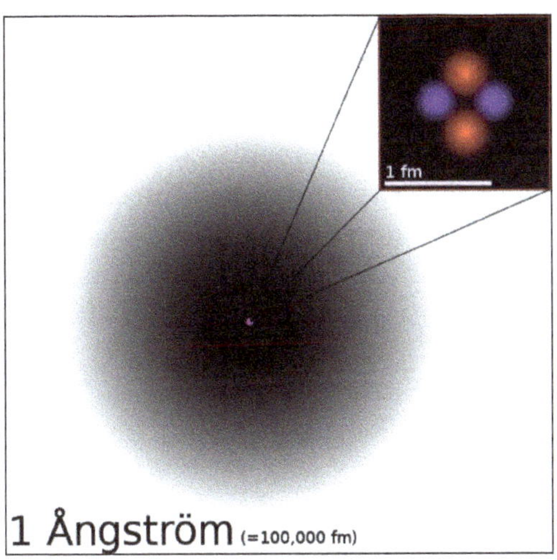

- Each chemical element is composed of extremely small particles that are indivisible and cannot be seen by the naked eye, called atoms. Atoms can neither be created nor destroyed. Pictured below is a helium atom. The purple and red dots represent the neutrons and protons in the nucleus. The black area around the nucleus represent the electron cloud.
- All atoms of an element are alike in mass and other properties, but the atoms of one element differ from all other elements. For example, gold and silver have different atomic masses and different properties.

- For each compound, different elements combine in a simple numerical ratio. The illustration below describes this rule. The second equation for the reaction is incorrect because half of an atom does not exist.

One Carbon atom + Four Hydrogen atoms → CH_4

~~½ Carbon atom + Two Hydrogen atoms → $C_{1/2}H_2$~~

Atomic theory can be used to answers the questions presented above. A pure gold necklace is made up of atoms. A pure gold necklace and a pure silver necklace are different because they have different atoms. Pure gold mixed with pure copper forms rose gold. The gold and copper atoms combine in a simple numerical ratio.

Dalton's theory has not proven to be correct under all circumstances. The first rule was proven incorrect when scientists divided atoms in a process called nuclear fission. The second rule was proven incorrect by the discovery that not all atoms of the same element have the same mass; there are different isotopes. However, these failures do not justify discarding the atomic theory. It correctly explains the law of conservation of mass: if atoms of an element are indestructible, then the same atom must be present after a chemical reaction as before and, and the mass must constant. Dalton's atomic theory also explains the law of constant composition: if all the atoms of an element are alike in mass and if atoms unite in fixed numerical ratios, the percent composition of a compound must have a unique value without regards to the sample analyzed. The atomic theory led to the creation of the law of multiple proportions.

Law of Multiple Proportions

The law of multiple proportions states that if two elements form more than one compound between them, the masses of one element combined with a fixed mass of the second element form in ratios of small integers. The illustration of the third rule of the atomic theory correctly depicts this law.

Discovering Electrons

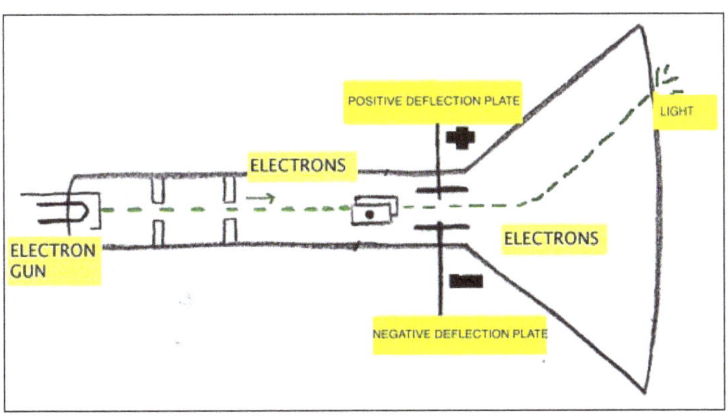

The first cathode-ray tube (CRT) was invented by Michael Faraday. Cathode rays are a type of radiation emitted by the negative terminal, the cathode, and were discovered by passing electricity through nearly-evacuated glass tubes. The radiation crosses the evacuated tube to the positive terminal, the anode. Cathode rays produced by the CRT are invisible and can only be detected by light emitted by the materials that they strike, called phosphors, painted at the end of the CRT to reveal the path of the cathode rays. These phosphors showed that cathode rays travel in straight lines and have properties independent of the cathode material (whether it is gold, silver, etc.) Another significant property of cathode rays is that they are deflected by magnetic and electric fields in a manner that is identical to negatively charged material. Due to these observations, J.J. Thompson concluded that cathode rays are negatively charged particles that are located in all atoms. It was George Stoney who first gave the term electrons to the cathode rays. The below figures depict the way that the cathode ray is effected by magnetics. The cathode ray is always attracted by the positive magnet and deflected by the negative magnets.

The Plum Pudding Model

After Thompson discovered the electron, he proposed the plum pudding model of an atom, which states that the electrons float in positively-charged material. This model was named after the plum-pudding dessert.

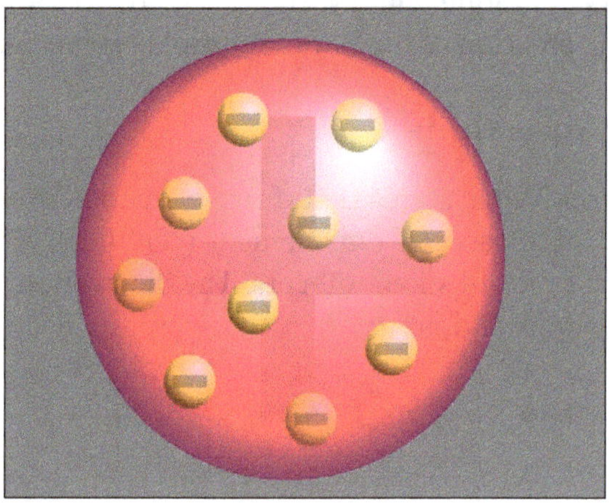

Discovery of the Proton

Ernest Rutherford performed a series of experiments studying the inner structure of atoms using alpha particles. Rutherford knew that alpha particles are significantly more massive than electrons and positively charged. Using the plum-pudding model for reference, Rutherford predicted that particles in an alpha beam would largely pass through matter unaffected, with a small number of particles slightly deflected. The particles would only be deflected if they happened to come into contact with electrons. According to the plum pudding model, this occurrence would be very unlikely. In order to test his hypothesis, Rutherford shot a beam of alpha particles at a thin piece of gold foil. Around the gold foil Rutherford placed sheets of zinc sulfide. These sheets produced a flash of light when struck by an alpha particle. However, this experiment produced results that contradicted Rutherford's hypothesis. Rutherford observed that the majority of the alpha particles went through the foil; however, some particles were slightly deflected, a small number were greatly deflected, and another small number were thrown back in nearly the direction from which they had come. Figure shows

Rutherford's prediction based off of the plum-pudding model (pink) and the observed large deflections of the alpha particles (gold).

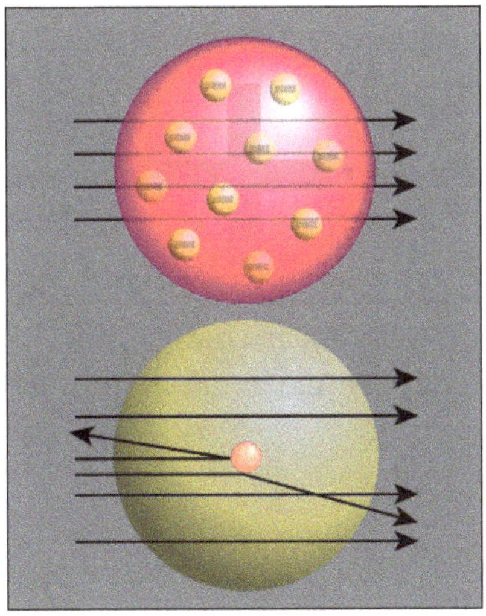

To account for these observations, Rutherford devised a model called the nuclear atom. In this model, the positive charge is held in an extremely small area called the nucleus, located in the middle of the atom. Outside of the nucleus the atom is largely composed of empty space. This model states that there were positive particles within the nucleus, but failed to define what these particles are. Rutherford discovered these particles in 1919, when he conducted an experiment that scattered alpha particles against nitrogen atoms. When the alpha particles and nitrogen atoms collided protons were released.

The Discovery of the Neutron

James Chadwick discovered a new type of radiation that consisted of neutral particles. It was discovered that these neutral atoms come from the nucleus of the atom. This last discovery completed the atomic model.

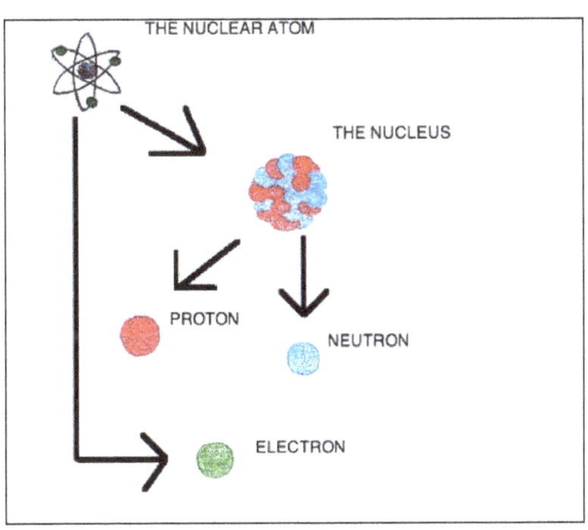

Molecular Orbital Theory

Molecular orbital theory is very simple to treat mathematically and is a landmark in molecular quantum mechanics. MOT was developed by F.Hund, R. S. Mulliken and J.E. Lennard-Jones. In MOT, the electrons do not belong to particular bonds but spread throughout the entire molecule, i.e., MOT approach looks at the molecule as a collection of nuclei and electrons. This theory is more developed than VBT.

Now, the question arises how to choose the MO trial wave-function. Molecular orbitals (MOs) in MOT are constructed by linear combination of atomic orbitals (LCAO) of atoms under consideration. For example, in an n-atom molecule, LCAO-MO approximation, the MO is written as,

$$\varphi_a = \sum_{i=1}^{n} c_i \psi_i$$

Let us say for n=2 (2 atom system), the MO becomes,

$$\varphi_a = c_1\psi_1 + c_2\psi_2$$

φ_a represents the approximated molecular orbital and ψ_1 and ψ_2 are atomic orbitals of the molecule under consideration.

Since c_1, c_2, ψ_1 and ψ_2 are all real, $\varphi_a^* = \varphi_a$

The approximate energy obtained using variation method is calculated using,

$$E_a = \frac{\int \varphi_a^* \hat{H} \varphi_a \, d\tau}{\int \varphi_a^* \varphi_a \, d\tau} \geq E_o$$

and as we already know that the variation method provides an upper bound to the ground state energy of the system E_o.

The above equation on simplification returns secular determinant,

$$\begin{vmatrix} H_{11} - E_a S_{11} & H_{12} - E_a S_{12} \\ H_{21} - E_a S_{21} & H_{22} - E_a S_{22} \end{vmatrix} = 0$$

The determinant in the above equation is called the secular determinant. The solution of this determinant gives a 2nd order secular equation as an approximation to ground state energy.

$$E_{a\pm} = \frac{H_{11} \pm H_{12}}{S_{11} \pm S_{12}}$$

Molecular Orbital Treatment of Hydrogen Molecule Ion

With this brief introduction, now let us take the case of hydrogen molecule ion. Hydrogen molecule ion H_2^+ has two nuclei and one electron. The Schrödinger equation for the system is,

$$\hat{H}\psi_{MO} = E\psi_{MO}$$

The total number of terms in Hamiltonian are,

Total number of terms in Hamiltonian $= \frac{(2+1)(2+1+1)}{2} = 6$

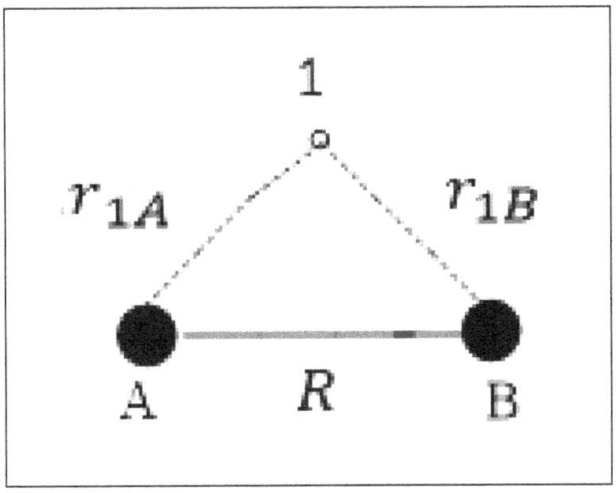

The Hamiltonian for H_2^+ is as follows:

$$\hat{H} = -\frac{h^2}{8\pi^2}\frac{1}{m_p}\nabla_A^2 - \frac{h^2}{8\pi^2}\frac{1}{m_p}\nabla_B^2 - \frac{h^2}{8\pi^2}\frac{1}{m_e}\nabla^2 + \frac{e^2}{4\pi\varepsilon_0 R} - \frac{e^2}{4\pi\varepsilon_0 r_1 A} - \frac{e^2}{4\pi\varepsilon_0 r_1 B}$$

Applying Born-Oppenheimer approximation (considering the nuclei fixed, the nuclear kinetic energy terms are neglected, $m_p > m_e$), we get electronic Hamiltonian as:

$$\widehat{H}_e = -\frac{h^2}{8\pi^2}\frac{1}{m_e}\nabla^2 + \frac{e^2}{4\pi\varepsilon_0 R} - \frac{e^2}{4\pi\varepsilon_0 r_1 A} - \frac{e^2}{4\pi\varepsilon_0 r_1 B}$$

where the first term in the kinetic energy of electrons, second term represents nucleusnucleus repulsion term and last two terms are electron and nucleus attraction terms respectively.

Further for hydrogen molecule ion, one can write,

$$\widehat{H}_e = \widehat{H}_e(H_{A1}) + X \quad \text{where } X = \frac{e^2}{4\pi\varepsilon_0 R} - \frac{e^2}{4\pi\varepsilon_0 r_1 B}$$

$$\widehat{H}_e = \widehat{H}_e(H_{B1}) + Y \quad \text{where } Y = \frac{e^2}{4\pi\varepsilon_0 R} - \frac{e^2}{4\pi\varepsilon_0 r_1 A}$$

where $\hat{H}_e(H_{A1})$ and $\hat{H}_e(H_{B1})$ represent exact electronic Hamiltonian for hydrogen atom A anf Hydrogen atom B.

$$\widehat{H}_e(H_{A1})\psi_{H_{A1}} = E_H\psi_{H_{A1}}$$

$$\widehat{H}_e(H_{B1})\psi_{H_{B_1}} = E_H\psi_{H_{B1}}$$

Now, for hydrogen molecule ion we need approximate molecular wave-function φ_a. We have one electron and two nuclei, so this electron can either be in the vicinity of nucleus A or B. So, one can guess φ_a.

$\varphi_a = \psi_{H_{A1}}$; when electron is in the vicinity of nucleus A.

$\varphi_a = \psi_{H_{B1}}$; when electron is in the vicinity of nucleus B.

There can be intermediate stages also, when the electron is equidistant from both A and B. Then, the real approximate wave-function will be the linear combination of both the above extreme cases;

$$\varphi_a(MO) = c_1\psi_{H_{A1}} + c_2\psi_{H_{B1}}.$$

$\varphi_a(MO)$ is the total approximate wave-function of MOT.

The values of coefficients c_1 and c_2 determine the contribution of the respective atomic orbitals of H_A and H_B.

Energy Expression for Hydrogen Molecule Ion

The energy for the system is given by equation,

$$E_{a\pm} = \frac{H_{AA} \pm H_{AB}}{S_{AA} \pm S_{AB}}$$

The secular determinant for the system becomes,

$$\begin{vmatrix} H_{AA} - E_aS_{AA} & H_{AB} - E_aS_{AB} \\ H_{BA} - E_aS_{BA} & H_{BB} - E_aS_{BB} \end{vmatrix} = 0$$

where,

$$H_{AA} = \int d\tau\psi_{H_{A1}}\widehat{H}_e\,\psi_{H_{A1}}$$

substituting,

$$\widehat{H}_e = \widehat{H}_e(H_{A1}) + X \qquad \text{where } X = \frac{e^2}{4\pi\varepsilon_0 R} - \frac{e^2}{4\pi\varepsilon_0 r_1 B}$$

$$H_{AA} = \int d\tau \psi_{HA1} \left[\widehat{H}_e(H_{A1}) + X \right] \psi_{HA1}$$
$$H_{AA} = \int d\tau \psi_{HA1} \left[\widehat{H}_e(H_{A1}) \right] \psi_{HA1} + \int d\tau \psi_{HA1} [X] \psi_{HA1}$$
$$H_{AA} = \int d\tau \psi_{HA1} [E_H] \psi_{HA1} + \int d\tau \psi_{HA1} [X] \psi_{HA1}$$
$$H_{AA} = [E_H] \int d\tau \psi_{HA1} + Q = E_H + Q$$

where $Q = \int d\tau \psi_{HA1} [X] \psi_{HA1}$;

S (Overlap integral) $= \int d\tau \psi_{HA1} \psi_{HA1} = 1$ as wave-function ψ_{HA1} is normalized.

$$H_{AB} = \int d\tau \psi_{HA1} \widehat{H}_e \psi_{HB1}$$

substituting,

$$\widehat{H}_e = \widehat{H}_e(H_{B1}) + Y \qquad \text{where } Y = \frac{e^2}{4\pi\varepsilon_0 R} - \frac{e^2}{4\pi\varepsilon_0 r_1 A}$$

$$H_{AB} = \int d\tau \psi_{HA1} \left[\widehat{H}_e(H_{B1}) + Y \right] \psi_{HB1}$$
$$H_{AB} = \int d\tau \psi_{HA1} \left[\widehat{H}_e(H_{B1}) \right] \psi_{HB1} + \int d\tau \psi_{HA1} [Y] \psi_{HB1}$$
$$H_{AB} = \int d\tau \psi_{HA1} [E_H] \psi_{HB1} + \int d\tau \psi_{HA1} [Y] \psi_{HB1}$$
$$H_{AB} = [E_H] \int d\tau \psi_{HA1} \psi_{HB1} + J$$

where $J = \int d\tau \psi_{HA1} [Y] \psi_{HB1}$;

$$S = \int d\tau \psi_{HA1} \psi_{HB1} = S$$
$$H_{AB} = E_H S + J$$

S(Overlap integral) $= \int d\tau \psi_{HA1} \psi_{HA1} = 1$ as wave-function as wave-functionis normalized

Substituting the values in energy expression gives,

$$E_{a\pm} = \frac{(E_H + Q) \pm (E_H S + J)}{1 \pm S}$$
$$E_{a+} = \frac{(E_H + Q) + (E_H S + J)}{1 + S} = E_H + \frac{(Q + J)}{(1 + S)}$$
$$E_{a-} = \frac{(E_H + Q) - (E_H S + J)}{1 - S} = E_H + \frac{(Q - J)}{(1 - S)}$$
$$E_{a\pm} = E_H + \frac{(Q \pm J)}{(1 \pm S)}$$

Coulomb Integral

$$Q=\int d\tau\psi_{H_{A1}}[X]\psi_{H_{A1}}$$

$$Q=\int d\tau\psi_{H_{A1}}\left[\frac{e^2}{4\pi\varepsilon_o R}-\frac{e^2}{4\pi\varepsilon_o r_{1B}}\right]\psi_{H_{A1}}$$

$$Q=\int d\tau\psi_{H_{A1}}\left[\frac{e^2}{4\pi\varepsilon_o R}\right]\psi_{H_{A1}}-\int d\tau\psi_{H_{A1}}\left[\frac{e^2}{4\pi\varepsilon_o r_{1B}}\right]\psi_{H_{A1}}$$

Since the internuclear distance in held constant, the first term in the above equation above can be simply written as,

$$\int d\tau\psi_{H_{A1}}\left[\frac{e^2}{4\pi\varepsilon_o R}\right]\psi_{H_{A1}}=\frac{e^2}{4\pi\varepsilon_o R}\int d\tau\psi_{H_{A1}}\psi_{H_{A1}}=\frac{e^2}{4\pi\varepsilon_o R}$$

The second term in the equation $Q=\int d\tau\psi_{H_{A1}}\left[\frac{e^2}{4\pi\varepsilon_o R}\right]\psi_{H_{A1}}-\int d\tau\psi_{H_{A1}}\left[\frac{e^2}{4\pi\varepsilon_o r_{1B}}\right]\psi_{H_{A1}}$ represents electrostatic attraction between nucleus B and an electron in 1s orbital centered on nucleus A and is called Coulomb integral. It represents electrostatic attraction and therefore, it has a negative value.

$$\int d\tau\psi_{H_{A1}}\left[\frac{e^2}{4\pi\varepsilon_o r_{1B}}\right]\psi_{H_{A1}}=Q_c\,(\text{Coulomb integral})$$

$$Q=\frac{e^2}{4\pi\varepsilon_o R}+Q_c$$

Exchange Integral

$$J=\int d\tau\psi_{H_{A1}}[Y]\psi_{H_{B1}}$$

$$J=\int d\tau\psi_{H_{A1}}\left[\frac{e^2}{4\pi\varepsilon_o R}-\frac{e^2}{4\pi\varepsilon_o r_{1A}}\right]\psi_{H_{B1}}$$

$$J=\int d\tau\psi_{H_{A1}}\left[\frac{e^2}{4\pi\varepsilon_o R}\right]\psi_{H_{B1}}-\int d\tau\psi_{H_{A1}}\left[\frac{e^2}{4\pi\varepsilon_o r_{1A}}\right]\psi_{H_{B1}}$$

Since the internuclear distance in held constant, the first term in the above equation above can be simply written as,

$$\int d\tau\psi_{H_{A1}}\left[\frac{e^2}{4\pi\varepsilon_o R}\right]\psi_{H_{B1}}=\frac{e^2}{4\pi\varepsilon_o R}\int d\tau\psi_{H_{A1}}=\frac{e^2}{4\pi\varepsilon_o R}S_{AB}=\frac{e^2}{4\pi\varepsilon_o R}S$$

The second term in equation $J=\int d\tau\psi_{H_{A1}}\left[\frac{e^2}{4\pi\varepsilon_0 R}\right]\psi_{H_{B1}}-\int d\tau\psi_{H_{A1}}\left[\frac{e^2}{4\pi\varepsilon_0 r_{1A}}\right]\psi_{H_{B1}}$ represents some kind of exchange of electron between nucleus A and nucleus B and is called exchange integral. The value of the integral depends on the physical overlap of the orbitals under consideration. This term has a negative value and its magnitude depends on the relative energies of the orbitals involved.

$$\int d\tau\psi_{H_{A1}}\left[\frac{e^2}{4\pi\varepsilon_0 r_{1A}}\right]\psi_{H_{B1}}=J_K$$

The value is maximum when the two orbitals have the same energy and it decreases as the difference of energies increases.

$$J=\frac{e^2}{4\pi\varepsilon_0 R}S+J_K$$

Substituting the values of Q and J in energy expression gives,

$$E_{a\pm}=E_H+\frac{\left(\left(\frac{e^2}{4\pi\varepsilon_0 R}+Q_c\right)\pm\left(\frac{e^2}{4\pi\varepsilon_0 R}S+J_K\right)\right)}{(1\pm S)}$$

$$E_{a+}=E_H+\frac{\left(\left(\frac{e^2}{4\pi\varepsilon_0 R}+Q_c\right)+\left(\frac{e^2}{4\pi\varepsilon_0 R}S+J_K\right)\right)}{(1+S)}$$

$$E_{a+}=E_H+\frac{e^2}{4\pi\varepsilon_0 R}+\frac{(Q_c+J_K)}{(1+S)}$$

And likewise we have,

$$E_{a-}=E_H+\frac{e^2}{4\pi\varepsilon_0 R}+\frac{(Q_c-J_K)}{(1-S)}$$

If for simplicity, we assume $S = 0$, the energy expression changes to

$$E_{a+}=E_H+\frac{e^2}{4\pi\varepsilon_0 R}+(Q_c+J_K)$$

$$E_{a-}=E_H+\frac{e^2}{4\pi\varepsilon_0 R}+(Q_c-J_K)$$

Since both Q and J are negative,

$$\Rightarrow E_{a+} < E_{a-}$$

For large internuclear distances, the terms Q_c and J_K vary in the following manner, J is inversely proportional to inter-nuclear distance R K varies inverse exponentially with R Since J has the form -1/R, it is largely cancelled by the nuclear repulsion term 1/R. Now K has a negative value which means,

$$E_{a+} < E_H \text{ and } E_{a-} > E_H$$

Molecular Wave-Function for Hydrogen Molecule Ion

We will now find the coefficients of the secular equation,

$$\begin{vmatrix} H_{AA} - E_a S_{AA} & H_{AB} - E_a S_{AB} \\ H_{BA} - E_a S_{BA} & H_{BB} - E_a S_{BB} \end{vmatrix} \begin{bmatrix} c_1 \\ c_2 \end{bmatrix} = 0$$

$$\begin{vmatrix} H_{AA} - E_a & H_{AB} - E_a S \\ H_{AB} - E_a S & H_{AA} - E_a \end{vmatrix} \begin{bmatrix} c_1 \\ c_2 \end{bmatrix} = 0$$

We get following equations as:

$$(H_{AA} - E_a) = \pm (H_{AB} - E_a S)$$

$$c_1 (H_{AA} - E_a) + c_2 (H_{AB} - E_a S) = 0$$

$$c_2 (H_{BA} - E_a S) + c_2 (H_{BB} - E_a S) = 0$$

From equations $(H_{AA} - E_a) = \pm (H_{AB} - E_a S)$ and $c_1 (H_{AA} - E_a) + c_2 (H_{AB} - E_a S) = 0$, we get

$$(c_1 + c_2)(H_{AB} - E_a S) = 0$$

$$\Rightarrow c_1 = c_2$$

From equations $(H_{AA} - E_a) = \pm (H_{AB} - E_a S)$ and $c_2 (H_{BA} - E_a S) + c_2 (H_{BB} - E_a S) = 0$, we get

$$(c_1 - c_2)(H_{AB} - E_a S) = 0$$
$$\Rightarrow c_1 = -c_2$$

$$\varphi_a (MO) = \psi_{H_{A1}} \pm \psi_{H_{B1}}$$

This means that the LCAO approach constructs two molecular orbitals viz., Bonding

Molecular Orbital (BMO) from constructive overlap of atomic orbitals (when $c_1 = c_2$) and Anti-Bonding Molecular Orbital (ABMO) from destructive overlap of atomic orbitals (when $c_1 = -c_2$).

Normalizing the above wave-function gives,

$$\int \varphi_a{}^* \varphi_a \, d\tau = 1$$

Let $|c_1| = |c_2| = c_+$

$$\int \varphi_a{}^* \varphi_a d\tau = |c_+|^2 \int (\psi_{H_{A1}} + \psi_{H_{B1}})^2 d\tau = 1$$

$$|c_+|^2 \int (\psi_{H_{A1}} + \psi_{H_{B1}})^2 d\tau = |c_+|^2 \int \psi_{H_{A1}}{}^2 + \psi_{H_{B1}}{}^2 + 2\psi_{H_{A1}}\psi_{H_{B1}} d\tau = |c_+|^2 (1+1+2S) = 1$$

$$|c_+|^2 (2+2S) = 1$$

$$|c_+|^2 = \frac{1}{2(1+S)}$$

$$|c_+| = \frac{1}{\sqrt{2(1+S)}}$$

and similarly with the negative combination,

$\int \varphi_a{}^* \varphi_a d\tau = |c_-|^2 \int (\psi_{H_{A1}} - \psi_{H_{B1}})^2 d\tau = 1$, we get

$$|c_-| = \frac{1}{\sqrt{2(1-S)}}$$

and with this the normalized BMO $(\varphi_{a+}(MO))$ and ABMO $(\varphi_{a-}(MO))$ becomes,

$$\varphi_{a+}(MO) = \frac{1}{\sqrt{2(1+S)}} (\psi_{H_{A1}} + \psi_{H_{B1}})$$

$$\varphi_{a-}(MO) = \frac{1}{\sqrt{2(1-S)}} (\psi_{H_{A1}} + \psi_{H_{B1}})$$

Molecular Orbital Treatment of Hydrogen Molecule Ion

We have already derived the energy and wave-function for hydrogen molecule ion using MOT approach. The results are as follows:

$$\varphi_{a+}(MO) = \frac{1}{\sqrt{2(1+S)}} (\psi_{H_{A1}} + \psi_{H_{B1}}) \text{ BMO}$$

$$\varphi_{a-}(MO) = \frac{1}{\sqrt{2(1-S)}}(\psi_{HA_1} + \psi_{HB_1})\ \text{ABMO}$$

$$E_{a+} = E_H + \frac{e^2}{4\pi\varepsilon_0 R} + (Q_c + J_K)$$

$$E_{a-} = E_H + \frac{e^2}{4\pi\varepsilon_0 R} + (Q_c - J_K)$$

Two H_2^+ MOs form - one bonding and one antibonding, from the AOs ψ_{HA_1} and ψ_{HB_1}.

Correlation Diagram

Now with the energy data and the molecular orbitals derived for hydrogen molecule ion, we can schematically represent it in the following manner.

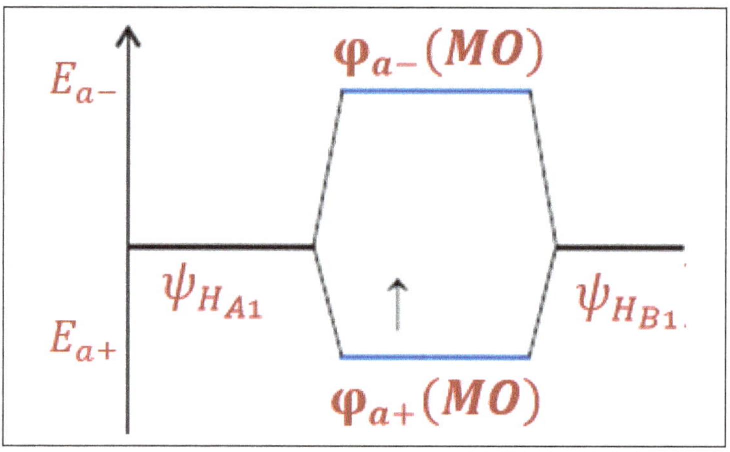

Molecular Orbital Energy Level diagram for H_2^+

The energy of BMO is less than that of the isolated atomic orbitals. Placing an electron in BMO decreases the energy of the system and contributes towards the formation of a chemical bond. Whereas the energy of ABMO is larger than that of the isolated atomic orbitals.

Energy Versus Internuclear Distance Plot

Another aspect of discussion is the variation of energy of MO with the internuclear distance. The plot for this variation is shown below.

As can be seen from the above plot, the energy curve corresponding to BMO contains a minimum (which corresponds to stable hydrogen molecule ion) at 1.06 Å (bond distance) with a depth of 2.8 eV (bond energy). The bond energy is the energy responsible for the formation of H_2^+. This is also referred to as stabilization energy or binding energy which is maximum at a distance of 1.06 Å (bond length of H_2^+ in case of BMO).

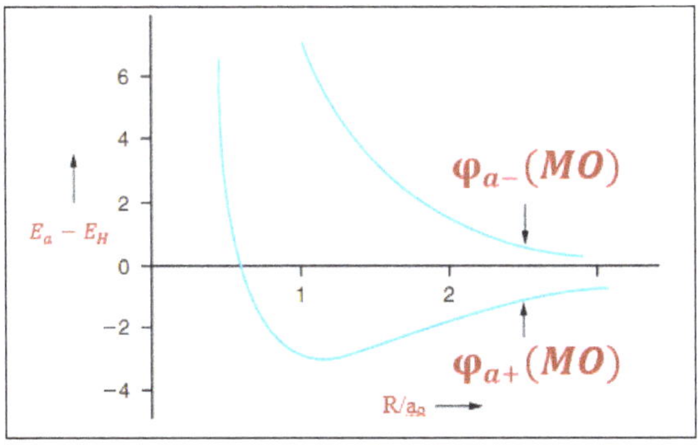

Plot of energies of the hydrogen molecule ion as a function of inter-nuclear distance.

Probability Densities Corresponding to MOs for $H_2^{\pm}$

The graphical representation of two wave-functions for hydrogen molecule ion (is shown below) can be obtained by simply taking the algebraic addition and subtraction of the two atomic orbitals,

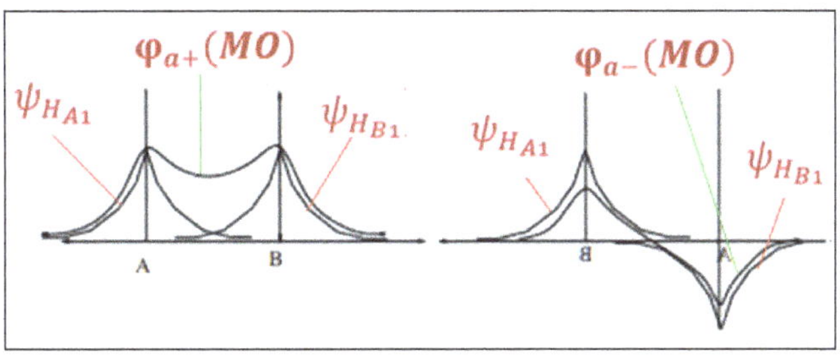

Graphical representation of MOs for H_2^+.

The probability density for the two wave-functions is obtained as:

$$\varphi_{a+}(MO)^2 = \frac{\psi_{HA1}^{\ 2} + \psi_{HB1}^{\ 2} + 2\psi_{HA1}\psi_{HB1}}{(2+2S)}$$

$$\varphi_{a-}(MO)^2 = \frac{\psi_{HA1}^{\ 2} + \psi_{HB1}^{\ 2} - 2\psi_{HA1}\psi_{HB1}}{(2-2S)}$$

The probability density plots are shown below:

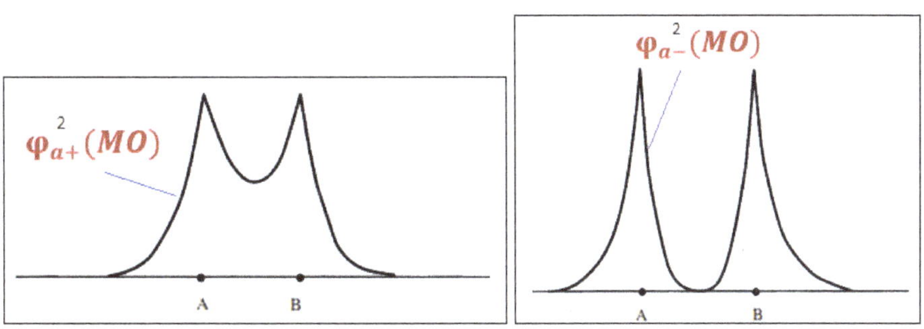

Probability densities corresponding to MOs for H_2^+.

One can see in the above plots that in case of ABMO, MO has zero value in a plane which is perpendicular to and bisects the axis between the two nuclei. This plane is called nodal plane. The probability density in this plane is also zero. That means, the probability of finding an electron on the nodal plane is zero and it is low in the region between the nuclei or in other words, the electron tends to avoid the region where it can interact with the nuclei. Whereas in case of BMO the probability of finding the electron between the two nuclei is relatively high.

Molecular Orbital Treatment of Hydrogen Molecule

Hydrogen molecule ion consists of two nuclei and one electron. Whereas hydrogen molecule is a two nuclei and two electron system.

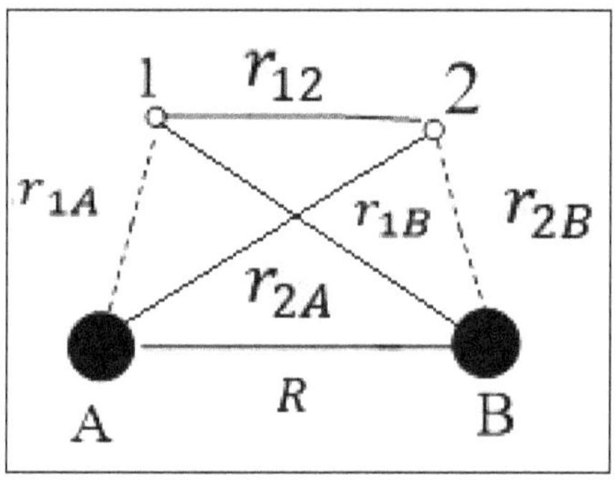

So, the total number of terms in Hamiltonian are:

$$\text{Total number of terms in Hamiltonian} = \frac{(2+2)(2+2+1)}{2} = 10$$

The Hamiltonian for H_2 molecule is as follows:

$$\hat{H} = -\frac{h^2}{8\pi^2}\frac{1}{m_p}\nabla_A^2 - \frac{h^2}{8\pi^2}\frac{1}{m_e}\nabla_1^2 - \frac{h^2}{8\pi^2}\frac{1}{m_e}\nabla_2^2 + \frac{e^2}{4\pi\varepsilon_0 R} - \frac{e^2}{4\pi\varepsilon_0 r_{1A}} - \frac{e^2}{4\pi\varepsilon_0 r_{1B}} - \frac{e^2}{4\pi\varepsilon_0 r_{2A}} - \frac{e^2}{4\pi\varepsilon_0 r_{2B}} + \frac{e^2}{4\pi\varepsilon_0 r_{12}}$$

Applying Born-Oppenheimer approximation (considering the nuclei fixed, the nuclear kinetic energy terms are neglected, $m_p > m_e$), we get,

$$\widehat{H_e} = -\frac{h^2}{8\pi^2}\frac{1}{m_e}\nabla_1^2 - \frac{h^2}{8\pi^2}\frac{1}{m_e}\nabla_2^2 - \frac{e^2}{4\pi\varepsilon_0 r_{1A}} - \frac{e^2}{4\pi\varepsilon_0 r_{1B}} - \frac{e^2}{4\pi\varepsilon_0 r_{2A}} - \frac{e^2}{4\pi\varepsilon_0 r_{1B}} + \frac{e^2}{4\pi\varepsilon_0 r_{12}} + \frac{e^2}{4\pi\varepsilon_0 R}$$

which can be re-written as,

$$\widehat{H_e} = \widehat{H_{1A}} + \widehat{H_{2B}} + \widehat{H'}$$

where $\widehat{H}_1$ and $\widehat{H}_2$ are the Hamiltonians of electron 1 in atom H_A and electron 2 in atom H_B respectively and $\widehat{H}'$ is the potential energy operator.

Molecular Wave-Function for Hydrogen Molecule

Now, for hydrogen molecule we need approximate molecular wave-function φ_a. There are two electrons electron and two nuclei. There are two atomic orbitals which combine to give two molecular orbitals similar to that in hydrogen molecule ion H_2^+ case but with one extra electron.

Similar to the case of hydrogen molecule ion, LCAO approach in case of hydrogen molecule constructs two molecular orbitals viz., Bonding Molecular Orbital (BMO) from constructive overlap of atomic orbitals and Anti-Bonding Molecular Orbital (ABMO) from destructive overlap of atomic orbitals.

MOs are of the same form as that of hydrogen molecule ion,

$$\varphi_a(MO) = c_1\psi_{H_A} + c_2\psi_{H_B}$$

And since both the nuclei are identical,

$$c_1 \pm c_2 = \frac{1}{\sqrt{2(1 \pm S)}}$$

$$\Rightarrow \varphi_a(MO) = \psi_{H_A} \pm \psi_{H_B}$$

The problem with the hydrogen molecule is that it is a two electron system where both the electrons are present in BMO. The correlation diagram for hydrogen molecule is given below:

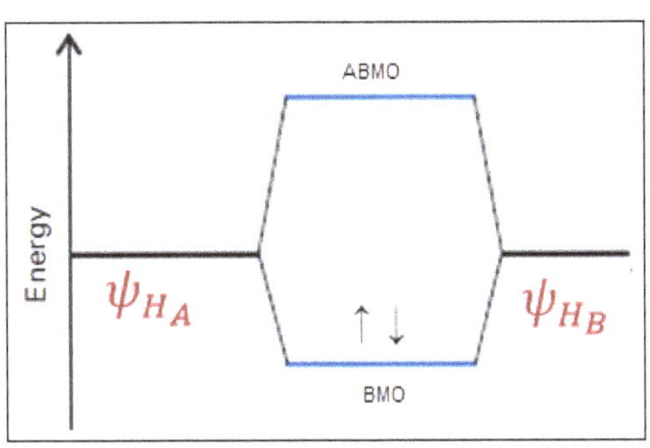

Molecular Orbital Energy Level diagram for H_2.

With this consideration, the total molecular wave-function for hydrogen molecule becomes,

$$\varphi_a(BMO) = (\psi_{H_A} + \psi_{H_B})_1(\psi_{H_A} + \psi_{H_B})_2$$

$$\varphi_a(BMO) = \psi_{H_{A1}}\psi_{H_{A2}} + \psi_{H_{A1}}\psi_{H_{B2}} + \psi_{H_{B1}}\psi_{H_{A1}} + \psi_{H_{B1}}\psi_{H_{A2}}$$

$\psi_{H_A1}\psi_{H_A2}$ and $\psi_{H_B1}\psi_{H_B2}$ represents ionic contribution as both the electrons 1 and 2 are present around H_a and H_b respectively. $\psi_{H_A1}\psi_{H_A2}$ represents a situation $H_A^- H_B^+$ and $\psi_{H_B1}\psi_{H_B2}$ 2 represents a situation $H_A^+ H_B^-$. Whereas, $\psi_{H_A1}\psi_{H_B2}$ and $\psi_{H_B1}\psi_{H_A2}$ represent covalent contributions. $\psi_{H_A1}\psi_{H_B2}$ represents a situation where electron 1 is around H_A and electron 2 around H_B. $\psi_{H_B1}\psi_{H_A2}$ represents a situation where electron 1 is around H_B and electron 2 around H_A.

Thus, the molecular wave-function for hydrogen molecule is a combination of ionic and covalent contributions. It has 50% ionic and 50% covalent contribution.

Energy Expression for Hydrogen Molecule

The energy for the system is determined in the following manner,

$$E_a = \frac{\int \varphi_a^*(BMO)\,\widehat{H}_e\,\varphi_a(BMO)d\tau}{\int \varphi_a^*(BMO)\,\varphi_a(BMO)d\tau}$$

Which reduced to $E_a = \int \varphi_a^*(BMO)\,\widehat{H}_e\,\varphi_a(BMO)d\tau$ as the molecular wave-function is normalized.

$$E_a = \int\left[\psi_{H_A1}\psi_{H_A2} + \psi_{H_A1}\psi_{H_B2} + \psi_{H_B1}\psi_{H_A2} + \psi_{H_B1}\psi_{H_B2}\right]$$
$$\widehat{H}_e\left[\psi_{H_A1}\psi_{H_A2} + \psi_{H_A1}\psi_{H_B2} + \psi_{H_B1}\psi_{H_A2} + \psi_{H_B1}\psi_{H_B2}\right]d\tau$$

$$\begin{aligned}
E_a = &\int \psi_{H_A1}\psi_{H_A2}\,\widehat{H}_e\,\psi_{H_A1}\psi_{H_A2}\,d\tau + \int \psi_{H_A1}\psi_{H_A2}\,\widehat{H}_e\,\psi_{H_A1}\psi_{H_B2}d\tau \\
&+ \int \psi_{H_A1}\psi_{H_A2}\,\widehat{H}_e\,\psi_{H_B1}\psi_{H_A2}\,d\tau + \int \psi_{H_A1}\psi_{H_A2}\,\widehat{H}_e\,\psi_{H_B1}\psi_{H_B2}d\tau \\
&+ \int \psi_{H_A1}\psi_{H_B2}\,\widehat{H}_e\,\psi_{H_A1}\psi_{H_A2}\,d\tau + \int \psi_{H_A1}\psi_{H_B2}\,\widehat{H}_e\,\psi_{H_A1}\psi_{H_B2}d\tau \\
&+ \int \psi_{H_A1}\psi_{H_B2}\,\widehat{H}_e\,\psi_{H_B1}\psi_{H_A2}\,d\tau + \int \psi_{H_A1}\psi_{H_B2}\,\widehat{H}_e\,\psi_{H_B1}\psi_{H_B2}d\tau \\
&+ \int \psi_{H_B1}\psi_{H_A2}\,\widehat{H}_e\,\psi_{H_A1}\psi_{H_A2}\,d\tau + \int \psi_{H_B1}\psi_{H_A2}\,\widehat{H}_e\,\psi_{H_A1}\psi_{H_B2}d\tau \\
&+ \int \psi_{H_B1}\psi_{H_A2}\,\widehat{H}_e\,\psi_{H_B1}\psi_{H_A2}\,d\tau + \int \psi_{H_B1}\psi_{H_A2}\,\widehat{H}_e\,\psi_{H_B1}\psi_{H_B2}d\tau \\
&+ \int \psi_{H_B1}\psi_{H_B2}\,\widehat{H}_e\,\psi_{H_A1}\psi_{H_A2}\,d\tau + \int \psi_{H_B1}\psi_{H_B2}\,\widehat{H}_e\,\psi_{H_A1}\psi_{H_B2}d\tau \\
&+ \int \psi_{H_B1}\psi_{H_B2}\,\widehat{H}_e\,\psi_{H_B1}\psi_{H_A2}\,d\tau + \int \psi_{H_B1}\psi_{H_B2}\,\widehat{H}_e\,\psi_{H_B1}\,d\tau
\end{aligned}$$

Quantitative analysis of the above expression gives energy minimum of 31.9 eV at 74 pm. When this energy is subtracted from the total energy of the two isolated hydrogen atoms, we get bond energy equal to 4.75 eV which is larger than that of hydrogen molecule ion (2.8 eV). This further implies that the bond distance in hydrogen molecule is smaller than that of hydrogen molecule ion.

Number of bonds in a molcule (Bond Order)

$$= \frac{(\text{number of electrons in BMO}) - (\text{number of electrons in ABMO})}{2}$$

Total Molecular Wave-Function

Over here, we have considered molecular wave-function taking into account spatial co-ordinates, i.e., we have taken into consideration only spatial wave-function. To further improvise, according to Pauli's principle, the wave-functions must be anti-symmetric with respect to exchange of the two electrons. For this, we need to consider spin wave-function. Spin refers to the spin angular momentum of the electron. The magnitude of spin is fixed.

For any multi-electron system, the total wave-function (electronic) has to be product of the spatial wave-function and spin wave-function.

$$\psi_e = \varphi_a(\text{MO}) = \psi_{\text{spatial}} \cdot \psi_{\text{spin}}$$

This product has to be antisymmetric with respect to the exchange of two electrons.

The spatial wave-function is symmetric,

$$\psi_{\text{spatial}} = \varphi_a(\text{BMO}) = \left(\psi_{H_A} + \psi_{H_B}\right)_1 \left(\psi_{H_A} + \psi_{H_B}\right)_2$$

For a two electron system, there are four possible spin combinations $\left(s = \frac{1}{2}(\alpha) \text{or} -\frac{1}{2}(\beta)\right)$,

$$\alpha(1)\ \alpha(2)$$
$$\beta(1)\ \beta(2)$$
$$(\alpha(1)\beta(2))$$
$$(\alpha(2)\beta(1))$$

The first two spin states represent states with parallel spins and the other two states represent states with anti-parallel spins. Since the fact that the electrons are fundamentally indistinguishable forces to consider the two linear combinations:

$$(\alpha(1)\ \beta(2) \pm \alpha(2)\beta(1))$$

Out of all the spin combinations, the symmetric ones are:

$$\psi_{\text{spin}} = \alpha(1)\alpha(2)$$
$$\psi_{\text{spin}} = \beta(1)\ \beta(2)$$

$$\psi_{spin} = \frac{1}{\sqrt{2}}\left(\alpha(1)\beta(2)+\alpha(2)\beta(1)\right)$$

and the anti-symmetric is,

$$\psi_{spin} = \frac{1}{\sqrt{2}}\left(\alpha(1)\beta(2)-\alpha(2)\beta(1)\right)$$

The criteria for assigning symmetry is: On exchanging electron (1) with electron (2)

- If the sign changes: Wave-function anti-symmetric.
- If the sign does not change: Wave-function symmetric.

For a two electron system, the complete wave-function can be obtained by multiplying orbital wave-function with spin function such that the complete wave-function is anti-symmetric under the interchange of orbital and spin co-ordinates.

Since the spatial wave-function is symmetric, the only possibility to make complete wavefunction anti-symmetric is to consider anti-symmetric spin wave-function,

$$\psi_e = \varphi_a(BMO) = \left[\left(\psi_{H_A}+\psi_{H_B}\right)_1\left(\psi_{H_A}+\psi_{H_B}\right)_2\right]\left[\frac{1}{\sqrt{2}}\left(\alpha(1)\beta(2)-\alpha(2)\beta(1)\right)\right]$$

The number of spin states associated with any wave-function determines the multiplicity of that state.

Number of spin states = 2S+1 (where S is the total spin of the state)

The above wave-function represents the ground state where spatial wave-function is symmetric and spin wave-function is anti-symmetric. This state is a singlet state (S = 0, 2S + 1 = 0). When the two electrons enter a single molecular orbital, they do so with paired spins (↑↓). Spin-pairing refers to the way in which electrons must arrange themselves in order to pack into the lowest energy orbital.

Transition-State Theory

Transition-state theory is also called activated-complex theory or theory of absolute reaction rates, treatment of chemical reactions and other processes that regards them as proceeding by a continuous change in the relative positions and potential energies of the constituent atoms and molecules. On the reaction path between the initial and final arrangements of atoms or molecules, there exists an intermediate configuration at which the potential energy has a maximum value. The configuration corresponding to this maximum is known as the activated complex, and its state is referred to as the transition state. The difference between the energies of the transition and the initial states is closely related to the experimental activation energy for the reaction; it represents the minimum energy that a reacting or flowing system must acquire for the transformation to take place. In transition-state theory, the activated complex is considered to have been formed in a state of equilibrium with the atoms or molecules in the initial state, and therefore its statistical and thermodynamic properties can be specified. The rate at which the final state is attained is determined by the number of activated complexes formed and the frequency with which they go over to the final state. These quantities may be calculated for simple systems by using statistical-mechanical principles. In this way the rate constant of a chemical or physical process may be expressed in terms of atomic and molecular dimensions, atomic masses, and interatomic or intermolecular forces. Transition-state theory can also be formulated in thermodynamic terms.

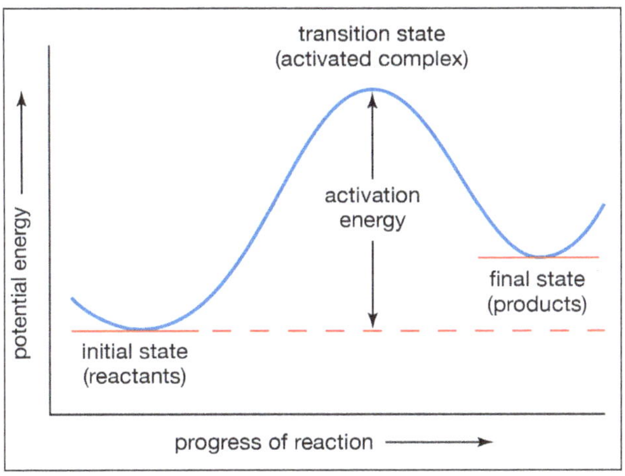

Potential-energy curve.

The activation energy represents the minimum amount of energy required to transform reactants into products in a chemical reaction. The value of the activation energy is equivalent to the difference in potential energy between particles in an intermediate configuration (known as the transition state, or activated complex) and particles of reactants in their initial state. The activation energy thus can be visualized as a barrier that must be overcome by reactants before products can be formed.

Activated state is then formed as highly energized and then unstable intermediate, which has transient existence and decomposes to form the products of reaction. The rate of reaction is given by the definite rate of decomposition of activated complex. The other most important assumption of STS theory is the existence of chemical equilibrium (quasi-equilibrium) between reactants and activated state (complex). Despite the fact that this equilibrium is different from classical chemical equilibrium, it can be described by the same thermodynamic treatment. It can be then summarized that TST theory assumes that chemical reaction (or any other activated process[2] proceeds as two-stage process.

Formation of Activated Complex

Decomposition of activated complex into the products of reaction as is shown in example in Figure. (The representation of energetic pathway derived from the intersection of the PES associated with the reaction.) (Since the catalysts provide an alternative reaction mechanism with lower E_a), the rate of reaction increases. This demonstration of significant effect of mechanism on the value of activation energy emphasizes the necessity to know the whole kinetic triplet (E_a, A and n) to complete the characterization of reaction kinetics.)

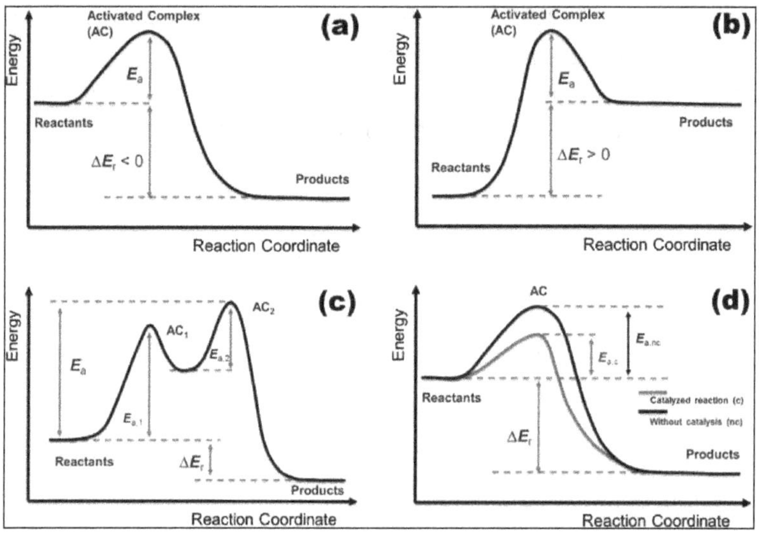

Potential energy diagram (potential energy profile or reaction coordinate diagram for exothermic (a) endothermic (b) multistep process (c) and catalyzed reaction (d).

A general stoichiometric equation for the chemical reaction between R_i chemical species (individuals, that is, atoms, molecules, ions, etc.) in the state marked by the appending (f):

$$\sum_i v_i R_i(f) = 0 \Rightarrow aA(f) + bB(f) + \ldots \Leftrightarrow sS(f) + tT(f) + \ldots$$

Where A, B... and S, T... are reactants and products (reaction products), respectively. Integers v_i = a, b, s, t... are the stoichiometric coefficients, i.e. the degree to which a chemical species participates in a reaction, that is conventionally negative for reactants and positive for products.

For the selected case of the three-atom bimolecular reaction of the type:

$$AB + C \leftrightarrow A + BC.$$

It is possible to write:

$$AB + C \leftrightarrow A \ldots B \ldots C \rightarrow A + BC;$$

where A···B···C is an activated complex. The energy difference between the energy of activated complex and reactants is termed as the activation energy (E_a).

Since the equilibrium of activated complex with reactant is assumed, it is possible to define the equilibrium constant of activation:

$$AB + C \leftrightarrow A...B...C;$$

as follows:

$$K^{\#} = \frac{\left[C^{\#}\right]}{[AB][C]} = \frac{\left[C^{\#}\right]}{\prod_i [C_i]^{v_i}};$$

where $\left[C^{\#}\right] = K^{\#}[AB][C]$ denotes the concentrations of activated complexes. The rate of the reaction is then given by the activated complex decomposition rate, i.e.:

$$A...B...C \rightarrow A + BC;$$

which is proportional to $\left[C^{\#}\right]$, so that the relation can be written:

$$r = v^{\#}\left[C^{\#}\right] = v^{\#}K^{\#}[AB][C];$$

where $v^{\#}$ is the activated complex decomposition frequency.

According to the statistical thermodynamics, the value of equilibrium constant in the reaction is given by the formula:

$$K^{\#} = \frac{Q_o^{\#}}{Q_{AB}Q_C}\exp\left(-\frac{E_o}{RT}\right);$$

where E_o is an energetic difference between the energy of activated state (product) and reactants that is given as relative to the energy of atoms in gaseous state at the temperature of 0 K. So in this particular case the value of $E_o = E_a$. The quantities $Q_o^{\#}$, Q_{AB}, and Q_C are the partition functions of activated state and reactant molecules.

Formulation of Partition Function

The partition function of molecule Q, which consists of multiple quantum microstates j (index for the microstate of the system) which share the same energy ε_j, i.e. the energy levels of the system, which are degenerated, can be expressed by the following formula:

$$Q = \sum_j g_j \exp\left(-\frac{\varepsilon_j}{k_B T}\right) = \sum_j g_j \exp\left(-\beta\varepsilon_j\right);$$

where g_j is the number of allowed equimolar quantum microstates (degeneracy factor) and k_B is the Boltzmann constant. The value of thermodynamic beta is defined as:

$$\beta = \frac{1}{k_B T} = \frac{1}{\tau} = \frac{1}{k_B}\frac{dS}{dE} = \frac{\partial \ln g_j}{\partial \varepsilon_j};$$

and the exponential factor exp $(-\beta\varepsilon_j)$ is known as the Boltzmann factor, S is the entropy and E is the energy. τ in Equation above is termed as the fundamental temperature that has a unit of energy, i.e. the joule $(J = kg \bullet m^2 / s^2)$. The denominator $k_B T$, i.e. the product of the Boltzmann's constant and thermodynamic temperature, that expresses energy on molecular scale, is also termed as the constant of distribution.

The partition function is then the summation of terms $\exp(-\varepsilon_j / k_B T)$ for all allowed quantum states (j) of the system, which can be considered for the sum of energy of independent moves, such as:

- Translation: The translational energy for one-dimensional motion is usually given by the following equation:

$$Q_{trans} = \frac{1}{2} m\overline{v}^2 = \frac{\overline{p}^2}{2m};$$

where m is the mass and $\overline{p}$ is the momentum of particle.

- Rotation: The rotational (angular) kinetic energy around the axis of rotation is given by the formula:

$$Q_{rot} = \frac{1}{2} I\omega^2;$$

where ω is the angular velocity and I is the moment of inertia around the axis of rotation. The comparison of equations $Q_{trans} = \frac{1}{2} m\overline{v}^2 = \frac{\overline{p}^2}{2m}$; and $Q_{rot} = \frac{1}{2} I\omega^2$; shows that the moment of inertia in rotating system is the parameter analogical to mass and the angular velocity is a quantity, which is analogical to the velocity.

- Vibration: The reactants, e.g, the species AB, C in equations AB + C $\leftrightarrow$ A + BC. – AB + C $\leftrightarrow$ A...B...C;, have 3 N-6 or 3 N-5 vibrational degrees of freedom (nVDF, please refer also to footnote 16) as they are non-linear or linear, respectively. The same can also be extended to the activated complex (A···B···C) that possesses $3 \bullet (N_{AB} + N_C) - 6$ or $3 \bullet (N_{AB} + N_C) - 5$ vibrational degrees of freedom if it is non-linear or linear, respectively.

- Energy of the electron states.

So it is possible to write the formula:

$$Q = Q_{trans}\, Q_{rot}\, Q_{vibr}\, Q_{el}.$$

The calculation of this energy then requires the information about allowed energetic levels and their degeneration in the molecule, which can be determined via the methods of quantum mechanics from the structure of this molecule, i.e., the molecular parameters.

The activated complex moving along one-dimensional reaction pathway, which is known as the reaction coordinate (collective variable), is decomposed to the product of the reaction. That enables to split the partition function into two parts:

- Contribution (one particular degree of freedom) that corresponds to the shift (vibration) along the reaction coordinate (Q_{rk}), which leads to the decomposition of activated complex.
- Contribution corresponding to another molecular degree of freedom (Q_{oth}); so it can be written:

$$Q^{\#} = Q_{rk}\, Q_{oth}.$$

The allowed energetic state for quantum oscillator vibrating along the reaction coordinate results from the solution of Schrödinger equation:

$$\varepsilon_n = \frac{nhv}{k_B T};$$

where n = 0, 1, ... ∞, v is the oscillator frequency and h is the Planck constant. The Q_{rk} partition function can be then expressed by the formula:

$$Q_{rk} = \sum_{n=0}^{\infty} \exp\left(-\frac{nhv}{k_B T}\right).$$

Using the following substitution:

$$\frac{nhv}{k_B T} = x;$$

In order to solve this geometric series, the relation can be written:

$$Q_{rk} = \frac{1}{1 - \exp\left(-\frac{hv}{k_B T}\right)}.$$

A small value of v means that the product $hv \gg k_BT$ and the denominator in Equation above should be developed in every convergent series (expansion or polynomial):

$$\exp\left(-\frac{hv}{k_BT}\right) = 1 - \frac{hv}{k_BT} + \frac{1}{2}\left(\frac{hv}{k_BT}\right)^2 - \frac{1}{6}\left(\frac{hv}{k_BT}\right)^3 + \frac{1}{24}\left(\frac{hv}{k_BT}\right)^4 - \cdots = 1 - \frac{hv}{k_BT}$$

It is then possible to write that:

$$Q_{rk} \approx \frac{1}{1 - 1 + \frac{hv}{k_BT}} = \frac{k_BT}{hv}.$$

Since the frequency of this vibration means the frequency of decomposition of activated state into the product $(v^{\#})$, the combination with Equation $K^{\#} = \frac{[C^{\#}]}{[AB][C]} = \frac{[C^{\#}]}{\prod_i [C_i]^{v_i}}$;

and substitution to Equation $r = v^{\#}[C^{\#}] = v^{\#}K^{\#}[AB][C]$; enable to formulate the rate of reaction as follows:

$$r = v^{\#}[C^{\#}] = v^{\#}K^{\#}[AB][C] =$$

$$v\# \frac{kT}{hv^{\#}} \frac{Q_{oth}}{Q_{AB}Q_C} \exp\left(-\frac{E_a}{RT}\right)[AB][C];$$

and then,

$$r = \frac{k_BT}{h} - \frac{Q_{oth}}{Q_{AB}Q_C} \exp\left(-\frac{E_a}{RT}\right)[AB][C].$$

Calculation of the Reaction Rate

Since the kinetic equation for the reaction can be written as follows:

$$r = k[AB][C];$$

where k is the rate constant, the combination of equations $K^{\#} = \frac{Q_o^{\#}}{Q_{AB}Q_C} \exp\left(-\frac{E_o}{RT}\right)$; and $r = \frac{k_BT}{h} - \frac{Q_{oth}}{Q_{AB}Q_C} \exp\left(-\frac{E_a}{RT}\right)[AB][C]$. then leads to the formula:

$$k = \frac{k_BT}{h} \frac{Q_{oth}}{Q_{AB}Q_C} \exp\left(-\frac{E_a}{RT}\right);$$

which enables to calculate the reaction rate from the properties of activated complex and reactants. Unfortunately, the evaluation of partition function of the activated state is usually difficult due to the lack of information about its structure. Therefore, the following formula is often used:

$$k = \frac{k_B T}{h} K^+;$$

where K^+ is the equilibrium constant of activation, which does not include particular part of partition function that corresponds to the vibration along the reaction coordinate. Assuming the parity $K^+ = K^\#$, the following relation can be written:

$$-RT \ln K^\# = \Delta G^\# = \Delta H^\# - T\Delta S^\#$$

The combination of above introduced equations $k = \frac{k_B T}{h} \frac{Q_{oth}}{Q_{AB} Q_C} \exp\left(-\frac{E_a}{RT}\right)$; - $-RT \ln K^\# = \Delta G^\# = \Delta H^\# - T\Delta S^\#$ then leads to the expression of general form of so-called Eyring-Polanyi equation:

$$k = \frac{k_B T}{h} \exp\left(\frac{\Delta S^\#}{R}\right) \exp\left(-\frac{\Delta H^\#}{RT}\right) = \frac{k_B T}{h} \exp\left(\frac{\Delta G^\#}{RT}\right) = v^\# K^\#;$$

where term $v^\# = k_B T / h$ is known as the universal frequency (frequency factor). The combination of Equation above with universal frequency of activated complex enables to solve the relation between the Arrhenius and Eyring rate constants as follows:

$$K_{Arrhenius} = A \bullet \exp(-E_a / RT) = k_{Eyring} / e = \left(v^\# K^\#\right) / e \Rightarrow k_{Arrhenius} / k_{Eyring} = 1/e.$$

Since every vibration does not necessarily result in the decomposition of activated complex into the products, the proportionality constant (transmission coefficient) κ can be introduced into the Eyring equation to get the relation:

$$k = \kappa \frac{k_B T}{h} \exp\left(\frac{\Delta S^\#}{R}\right) \exp\left(-\frac{\Delta H^\#}{RT}\right) = \kappa \frac{k_B T}{h} \exp\left(-\frac{\Delta G^\#}{RT}\right) = \kappa v^\# K^\# = k^\# K^\#$$

where $k^\#$ is assumed to be directly proportional to the frequency of the vibrational mode, which is responsible for the decomposition of the activated complex into products. Furthermore, the transmission coefficient (κ) is in most cases considered to be close to unity, so Equation above comes to Eq. $K_{Arrhenius} = A \bullet \exp(-E_a / RT) = k_{Eyring} / e = \left(v^\# K^\#\right) / e \Rightarrow k_{Arrhenius} / k_{Eyring} = 1/e$. Since $k^\# \geq k$ the value of transmission coefficient is then $0 \leq \kappa \leq 1$.

The temperature dependence of the rate constant k is then given by the Arrhenius law.

The comparison of this equation to Arrhenius enables to derive that for the processes occurring in the condensed phases $E_a \approx \Delta H^{\#}$:

$$\Delta H^{\#} = E_a - RT;$$

and the frequency factor can be considered as independent on temperature. The entropy of activation can be calculated from the relation:

$$\Delta S^{\#} = R\ \text{In}\left(\frac{hA}{k_B T}\right);$$

and:

$$\Delta G^{\#} = \Delta H^{\#} - T\Delta S^{\#}.$$

The linearized (logarithmic) form of euation $k = \frac{k_B T}{h}\exp\left(\frac{\Delta S^{\#}}{R}\right)\exp\left(-\frac{\Delta H^{\#}}{RT}\right) = \frac{k_B T}{h}\exp\left(\frac{\Delta G^{\#}}{RT}\right) = v^{\#}K^{\#};$, which can be written as follows:

$$\text{In}\left(\frac{k}{T}\right) = -\frac{\Delta H^{\#}}{R}\frac{1}{T} + \text{In}\left(\frac{k_B}{h}\right) + \frac{\Delta S^{\#}}{R};$$

enables to determine the enthalpy ($\Delta H^{\#}$) and the entropy of activation ($\Delta S^{\#}$) from the investigation of reaction rate (kinetics) of reaction at different temperatures. The plot (so-called Eyring plot) of ln (k/T) versus reciprocal temperature (1/T) provides straight line with the slope of $-\Delta H^{\#}/R$ and the intercept $\ln(k_B/h) + \Delta S^{\#}/R$.

The examples of Arrhenius (a) and Eyring plot (b) for the process of thermal decomposition of calcite, i.e. the trigonal polymorph of $CaCO_3$, in an inert atmosphere of N_2:

$CaCO_3(s_1) \xleftrightarrow{T} CaO(s_2) + CO_2(g)$; are introduced in figure.

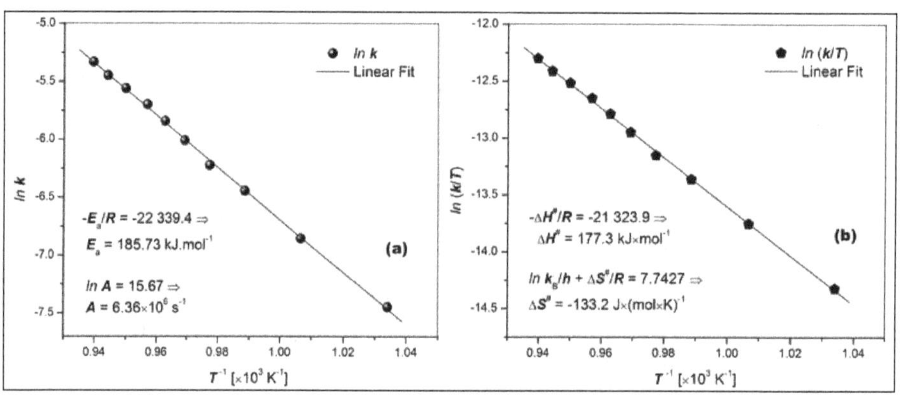

An example of calculation of enthalpy and entropy of activation for the process of thermal decomposition of calcite: the temperature dependence of the rate constant presented in the form of Arrhenius plot (a) and Eyring plot (b).

The slope of Arrhenius plot is used to write the relation for the activation energy (Arrhenius activation energy) as follows:

$$E_a = -R\left(\frac{\partial \ln k(T)}{\partial(1/T)}\right)_p = -p\left(\frac{\partial \ln k(T)}{\partial(1/V_m(T))}\right)_p;$$

where V(T) is the volume of ideal gas at temperature T. That means that:

$$\ln k(T) = \ln A - \frac{E_a}{pV_m(T)} \Rightarrow k(T) = \exp\left(-\frac{E_a}{pV_m(T)}\right).$$

The isobaric character of the process means that the relation between equations $Q_{rk} \simeq \frac{1}{1-1+\frac{h\nu}{k_B T}} = \frac{k_B T}{h\nu}$, $E_a = -R\left(\frac{\partial \ln k(T)}{\partial(1/T)}\right)_p = -p\left(\frac{\partial \ln k(T)}{\partial(1/V_m(T))}\right)_p$; and $\ln k(T)$ $\ln A - \frac{E_a}{pV_m(T)} \Rightarrow k(T) = \exp\left(-\frac{E_a}{pV_m(T)}\right)$ is given by the Charles's Law:

$$\frac{V_m}{T} = \text{const.} = \frac{R}{p};$$

which is also known as the law of volumes.

Frequency Factor as the Function of Temperature

The temperature dependence for experimental (marked by subscript exp), i.e. usual form of the Arrhenius equation, and theoretically calculated (th) reaction rate constants can be written as follows:

$$k_{exp} = A_{exp} \exp\left(-\frac{E_{a,exp}}{RT}\right);$$

and,

$$k_{th} = B(T)\exp\left(-\frac{W}{RT}\right);$$

where the pre-exponential factor B(T) is the function of temperature and W is the difference between energy of reactants and activated complex in the basic state. According to the transition state theory, the value of B(T) is approximately proportional to the temperature.

The equations for the experimental and calculated (theoretical) activation energy can be then written as follows:

$$E_{a,exp} = RT^2 \frac{d \, In k_{exp}}{dT};$$

and,

$$E_{a,exp} = RT^2 \frac{d \, In k_{exp}}{dT} = RT^2 \frac{d \, In \, B(T)}{dT} + W = \Theta \, RT + W;$$

where,

$$\Theta = T \frac{d \, In \, B(T)}{dT}.$$

The substitution to Equation $k_{th} = B(T)\exp\left(-\frac{W}{RT}\right)$; then leads to the formula:

$$k_{th} = B(T)\exp(\Theta)\exp\left(-\frac{E_{th}}{RT}\right).$$

In the case that,

$$B(T) = CT;$$

where C is a constant, it can be written that,

$$\Theta = T \frac{d \, In \, B(T)}{dT} = 1;$$

and then,

$$k_{th} = B(T)\exp\left(-\frac{E_{th}}{RT}\right).$$

That also means the parity:

$$E_{th} = E_{exp};$$

and,

$$A_{th} = B(T) = CT.$$

Since the value of Θ is close to one, the pre-exponential (frequency) factor can be considered as independent of temperature.

Limitations of Transition State Theory (TST)

The universality of the transition state theory provides a conceptual foundation for understanding the course and rate of elementary reaction on molecular scale. On the other hand, the TST theory can fail in some cases.

- If applied to each elementary step of multistep reactions.
- When the intermediates are very short-living, so that the Boltzmann distribution of energies is not reached before the process continues to the next step.
- The transition state theory also fails for some reactions at high temperatures due to the more complex motions of molecules or at very low temperatures due to the quantum tunneling.

Furthermore, the TST is also based on the assumption that atomic nuclei behave according to the classic mechanics. It is assumed that unless atoms or molecules collide with enough energy to form the transition structure (activated complex), the reaction does not occur. However, according to the quantum mechanics, for any barrier with a finite amount of energy, there is a possibility that particles can still tunnel across the barrier.

Kinetic Isotope Effect

The discovery of isotopes of the elements resulted from a thorough study of the properties of radioactive substances during the first years of the twentieth century.

> "The word isotope signifies 'the same place' in allusion to isotopes occupying the same place in the Periodic Table".

The conclusion that certain species of atom have identical chemical properties but different radioactive properties comes from chemical studies, which accompanied the study of radioactivity. The fundamental theory of the chemical isotope effect was established in 1947 by Urey and Bigeleisen and Mayer who stated:

> "It is pointed out that the possibility of chemical separation of isotopes is a quantum effect. This permits a direct calculation of the difference in the free energies of two isotopic molecules."

The kinetic isotope effect (KIE) is the effect of isotopic substitution on the reaction rate. Early research on the effect of the difference in mass between isotopes upon the rate of chemical transformation has been limited entirely to the very lightest elements, especially hydrogen; although the effect of isotopic mass upon the rate of many physical properties has been extend throughout the periodic table. The heavy atom kinetic isotope effect studies became possible after the Second World War with the availability of radioactive isotopes and mass spectrometers for the measurement of stable isotope ratios.

The origin of all isotope effects lies in the differences of various vibrational modes of a molecule which arise when one isotope is substituted with another. There are two kinds of KIE recognized:

1. Primary kinetics isotope effect, i.e. the effect of isotopic substitution on the rate constant of transformation of isotopologues. Isotopologues are molecules which differ only in their isotopic composition (the number of isotopic substitutions or also isotopic identities), for example, protium containing water (H_2O) and deuterium bearing semiheavy (HDO) or heavy water (D_2O). Other examples are CH_4 and CH_3D, or $^{12}C^{16}O_2$, $^{13}C^{16}O_2$, and $^{12}C^{18}O^{16}O$, and so on.

For the multiple isotopic substitution (two or more rare isotopes) in isotopologues, the name "clumped" isotopologues are used. If isomers have the same number of each isotopic atoms but they occupy different symmetrically non-equivalent positions, e.g. $^{14}N^{15}N^{16}O$ and $^{15}N^{14}N^{16}O$, they are termed as isotopomers, i.e. isotopic isomers.

For heavy atoms (heavy atoms isotope effects), the maximum reaction rate constant ration is proportional to the square root of inverse isotopic masses:

$$\frac{k_x}{k_y} = \sqrt{\frac{M_y}{M_x}}.$$

2. Secondary kinetics isotope effect (steric isotope effect), which is attributed to different vibrational amplitudes of isotopologues, e.g. the mean-square of amplitudes of vibration of C-H bonds is greater than that of C-D bonds.

The effect of isotopic mass on certain chemical equilibria was used to separate the isotopes. From this point of view, the classic theory (formulated in 1947), i.e. the mass-dependent theory, assumes that at a constant temperature, the isotope enrichment factor is proportional to the isotopic mass difference and inversely proportional to the product of masses of two isotopes. For as long as four decades after the formulation of the mass-dependent theory, a failure of the isotope effect in chemical exchange equilibria to follow the theory has not been reported. The mass-independent isotope fractionations were first observed for O and S in 1983.

Also, the course of crystallization could be affected by isotopic composition hence rapidly grown crystals could have a disequilibrium elemental and isotopic concentrations, which result from the diffusion limitation and diffusivity differences in the growth medium. The isotopic composition is also affected by the presence of simple organic molecules and could be a crystal-face-dependent. The variations in oxygen ($^{18}O/^{16}O$) and hydrogen ($^{2}H/^{1}H$) isotope ratios of water and ice are powerful tools in hydrology and ice core studies. The fractionation of Cu, Fe, and Zn isotopes during the oxidative weathering of sulfide-rich rocks could also have an impact on the distributions of stable Fe, Cu, and Zn isotopes in natural waters.

Historically, the isotope techniques came into use in catalysis in 1960s with the investigation of oxygen and other molecules reactivity. The study of dynamics of isotope transfer under steady-state reaction conditions, the so-called Steady-State Isotopic Transient Kinetic Analysis (SSITKA), is successfully employed in the investigation of mechanism of heterogeneous catalytic reactions and to estimate the surface concentrations of intermediates and reaction rate coefficients. SSITKA method was initially developed by Happel, Bennett, and Biloen.

"Proper understanding of heterogeneous catalysis is linked to understanding of transient effects. The interpretation requires a model based on elementary steps, and various steps may be rate-determining as the conversion and temperature vary."

The basic idea of the isotope dynamic experiment is as follows. When the reaction under study achieves the steady state, the initial feed gas flow is stepwise replaced by another flow differing in the isotope composition, but absolutely identical in a chemical one. The composition of feed gas and reaction products is continuously measured using a quadrupole mass spectrometer.

Marcus-Hush Theory

Marcus Hush theory (M-H theory) was developed in 1956 by Rudolph A. Marcus which explains the fundamentals of the redox/ electron transfer reactions in terms of the rate of jumping/moving an electron from oxidant species (electron donor) to the reductant (electron acceptor). M-H theory works as alternative Eyring's transition state theory for electron transfer reactions where bond making or bond breaking does not occur.

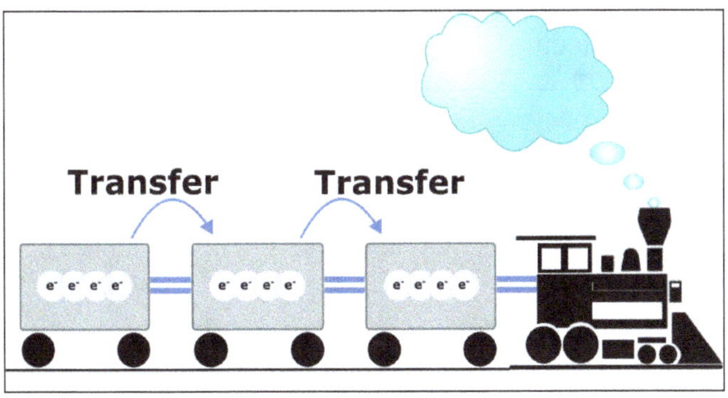

Vibration Barrier

According to M-H theory vibration barrier exists that hampers the redox/electron transfer reactions. The equilibrium structures for both the redox partners are dissimilar and the oxidation state to be changed during the course of reaction. For an electron transfer to take place, the electronic movement must be faster relative to that of nuclear movement. The vertical process feature suggests that nuclear locations do not change at the time of reaction. According to the energy conservation law, the transfer of electron can merely occur on the cut surface of two potential curves whose energy is equal to that of thermal activation energy. The structural reorganization of each complex takes place by bringing the distant ligand closer and the closer to distant by adjusting all the M-L bond lengths to be equal (λ_i inner reorganisation energy). This reorganization is associated to facilitate the electron transfer that requires both redox centres must have a alike structure during the course of electron transfer. As the vibration barrier is proportional to the distance between redox centres the closer attachment of both the redox partners is desired to make the redox barrier low and thus electron transfer to be faster.

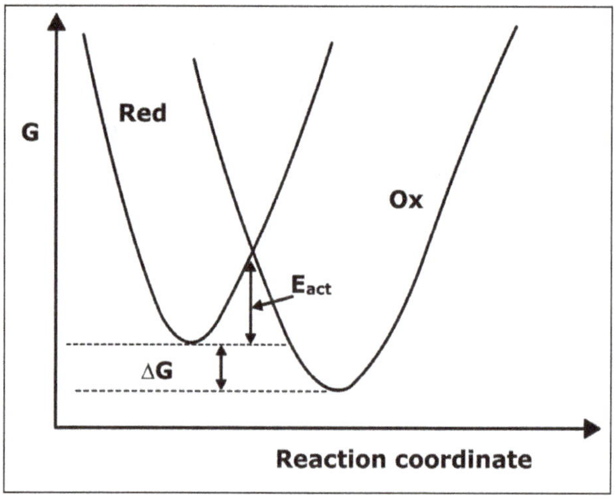

Solvation Barrier

Another factor that influences redox process is solvation barrier which is related to the organization of solvent molecules.

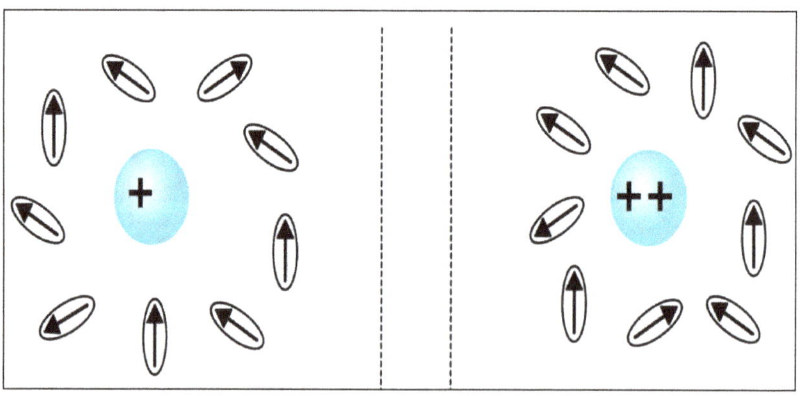

Electron transfer determined by solvent organization reorganization energy:

$$\lambda = \lambda_{in} + \lambda_{out}$$

Reorganization Energy

(a) Inner sphere reorganization energy.

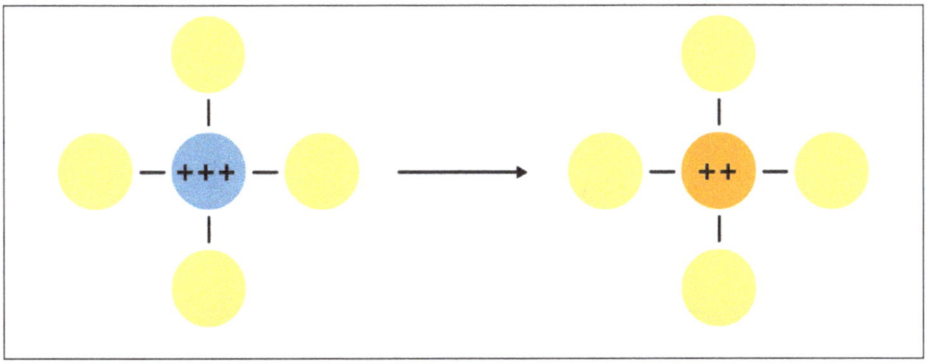

Inner sphere "breathing" of illustrated by following harmonic estimation:

$$\lambda_{in} = \frac{1}{2} m\omega^2 (q_f - q_i)^2$$

(b) Outer sphere reorganization energy.

Various polarization separations:

- Fast polarization: Electronic polarizability with a rate of 10^{-15}-10^{-16} sec.
- Slow polarization: Atomic action such as distortion and orientation.
- Total polarization: Fast polarization + slow polarization.

$$\overrightarrow{P} = \overrightarrow{P_s} + \overrightarrow{P_f}$$

The energy is required to change the solvent shell about the complexes at the time of absolute electron transfer reaction and this energy is associated with the outer reorganisation energy λ_o. The Marcus tale is principally based on the free energies (ΔG) of reactants and products which are roughly square functions of the reaction coordinate.

$$\Delta G^* = \frac{1}{4}\left[1 + \frac{\Delta G^\circ}{\lambda}\right] = \frac{[\lambda + \Delta G^\circ]^2}{4\lambda}$$

$\Delta G^* =$ free enthalpy $(-\text{ve})$;

$\Delta G^* =$ activation energy, $\lambda = \lambda_i + \lambda_o$

The reaction profiles can be discriminated by following three ways:

- $(\lambda + \Delta G^o) > 0$: Marcus's normal region, rate of reaction enhances with increase in ΔG^o.

- $(\lambda + \Delta G^o) = 0$: reaction rate reaches to maximum since there is no activation barrier for electron transfer to occur (diffusion controlled reaction).
- $(\lambda + \Delta G^o) < 0$: Marcus's inverted case and rate reaction diminishes since activation barrier increases.

Reactions as Example for Marcus Theory

$$\left[Co(bipy)_3\right]^{2+} + \left[Co^*(bipy)_3\right]^{3+} \xleftrightarrow{K_{11}} \left[Co(bipy)_3\right]^{3+} + \left[Co*(bipy)_3\right]^{2+}$$

Self exchange reaction-1

$$\left[Co(terpy)_2\right]^{2+} + \left[Co^*(terpy)_2\right]^{3+} \xleftrightarrow{K_{22}} \left[Co(bipy)_2\right]^{3+} + \left[Co*(terpy)_2\right]^{2+}$$

Self exchange reaction-1

$$\left[Co(bipy)_3\right]^{2+} + \left[Co(terpy)_2\right]^{3+} \xleftrightarrow{K_{12}} \left[Co(bipy)_3\right]^{3+} + \left[Co(terpy)_2\right]^{2+}$$

Cross reaction

* = isotope labelling

Marcus Theory of Electron Transfer: A Donor-Acceptor Interpretation

- Marcus theory from a molecular viewpoint is characteristically applicable to the outer sphere ET between an electron donor (D, reductant) and an electron acceptor (A, oxidant).
- Let us assume donor (D) and acceptor (A) as neutral molecules to facilitate the discussion so that electrostatic interactions/forces may be avoided.
- It is worth speculating and mentioning that either of donor or acceptor may lie in a photoexcited state. This, there is a possibility of photoinduced electron transfer (PET) process to occur.
- In principle, the concept of Marcus's model applies equally well for the electron transfer, both in the ground state and excited state except a change in the energies of starting stage.

A reaction involving homogenous mixture of donor (D) and acceptor (A), exhibits second-order kinetics and can be divided into three steps:

Precursor Complex Formation

Diffusion D and A having a rate constant k_a resulting in the formation of precursor complex D|A via an outer sphere mechanism. The precursor complex cleavage without electron transfer (ET) is illustrated by dissociation rate constant k_d.

$$D + A \underset{K_{-a}}{\overset{K_a}{\rightleftharpoons}} D/A$$

Successor Complex Formation

The restructuring of the precursor complex (D|A) endures to form a transition state which serves as a platform for feasible electron transfer (ET) to afford a successor complex D+|A–.

$$D/A \underset{K_{-ET}}{\overset{K_{ET}}{\rightleftharpoons}} D^+/A^-$$

The Franck Condon principle is obeyed right from the formation of precursor to successor complexes. Even the transition state must also be the same to form the successor complex.

Successor Complex Dissociation

Eventually, the successor complex dissociation takes place to separate cation (D^+) and anion (A^-).

$$D^+/A^- \xrightarrow{K_d} D^+ + A^-$$

Overall,

$$D + A \underset{K_{-a}}{\overset{K_a}{\rightleftharpoons}} D/A \underset{K_{-ET}}{\overset{K_{ET}}{\rightleftharpoons}} D^+/A^- \xrightarrow{K_d} D^+ + A^-$$

Application fo Marcus – Hush Theory in Bioinorganic Chemistry

Oxygen transport during respiration: Hemoglobin acts as “electron carrier” protein by carrying O_2 from lung to the cells in vertebrates.

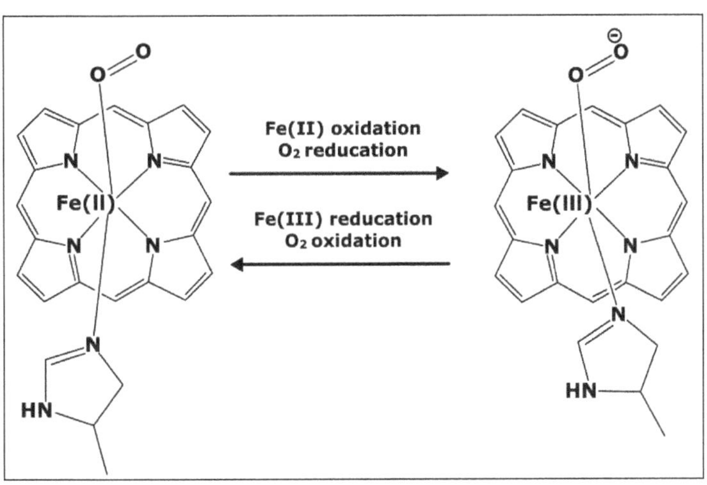

It is evident from above figure that the oxidation state of Fe(II) changes to Fe(III) and vice versa. Likewise, O_2 is reduced to O_2^- in forward step and in the backward step O_2^- is reduced to O_2.

Valence Bond Theory

Valence Bond Theory was the first quantum mechanical treatment to account for chemical bonding. This theory was first introduced by Heitler and London in 1927 and subsequently by Slater and Pauling in 1930s. It is sometimes referred to as Heitler-London-SlaterPauling (HLSP) theory.

VBT draws on the Lewis concept of a covalent bond as a shared pair of electrons (electron pair bond) i.e., valence bond approach is similar to the concept of valence bond between two atoms In this method, a molecule is considered as a collection of atoms and the interactions between different atoms are considered. With this, the question is how VBT chooses the trial/approximate wave-function.

Postulates of VBT

The main postulates of this theory are as follows:

- Overlapping of two half - filled valence atomic orbitals of two different atoms results in a covalent bond.
- The electrons in the overlapping orbitals get paired, localized and concentrated between the nuclei of two atoms.
- Only two electrons with their spins paired may be shared by one set of overlapping orbitals.
- The electron density between two bonded atoms increases due to overlapping. This confers stability to the molecule.
- The strength of the bond is directly proportional to the extent of overlap. Greater the extent of overlapping, stronger is the bond formed.
- The direction of the covalent bond is along the region of overlapping of the atomic orbitals i.e., covalent bond is directional.

VBT of Hydrogen Molecule

Before discussing the valence bond treatment for hydrogen molecule, it is important to note that VBT cannot solve Schrödinger equation for hydrogen molecule ion as VBT approximation considers overlapping of atomic orbitals each with one electron respectively. And hydrogen molecule ion contains only one electron. For this reason, hydrogen molecule ion cannot be explained by VBT.

Let us now consider the simplest case of chemical bond, i.e., the hydrogen molecule H2 which has two nuclei and two electrons.

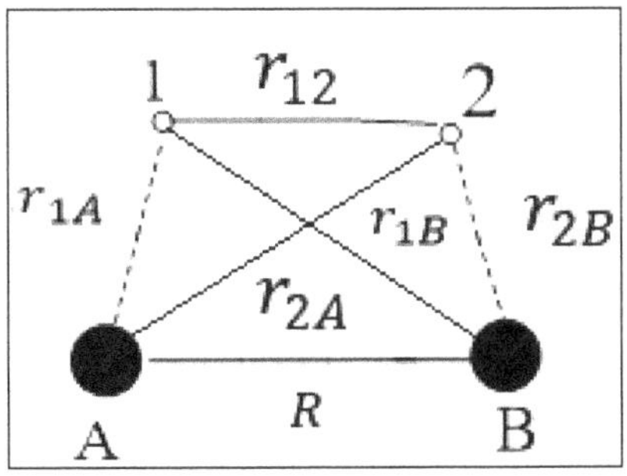

So, the total numbers of terms in Hamiltonian are:

$$\text{Total number of terms in Hamiltonian} = \frac{(2+2)(2+2+1)}{2} = 10$$

The Hamiltonian for H_2 molecule is as follows:

$$\hat{H} = -\frac{h^2}{8\pi^2}\frac{1}{m_p}\nabla_A^2 - \frac{h^2}{8\pi^2}\frac{1}{m_p}\nabla_B^2 - \frac{h^2}{8\pi^2}\frac{1}{m_e}\nabla_1^2 - \frac{h^2}{8\pi^2}\frac{1}{m_e}\nabla_2^2 + \frac{e^2}{4\pi\varepsilon_0 R} - \frac{e^2}{4\pi\varepsilon_0 r_{1A}} - \frac{e^2}{4\pi\varepsilon_0 r_{1B}} - \frac{e^2}{4\pi\varepsilon_0 r_{2A}}$$
$$- \frac{e^2}{4\pi\varepsilon_0 r_{2B}} + \frac{e^2}{4\pi\varepsilon_0 r_{12}}$$

which can be rewritten as,

$$\widehat{H_e} = \widehat{H_{A1}} + \widehat{H_{B1}} + \widehat{H'}$$

where $\widehat{H_{A1}}$ and $\widehat{H_{B2}}$ are the Hamiltonians of electron 1 in atom HA and electron 2 in atom HB respectively and $\widehat{H'}$ is the potential energy operator.

Wave-Function for Hydrogen Molecule

Since hydrogen molecule is a two nuclei and two electron system, there are two possibilities.

Case: If the atoms are widely spaced (with no interaction with each other), the electronic wave-function for the system can be written as,

$$\psi_{VB} = \psi_{H_A 1}\psi_{H_B 2}$$

where $\psi_{H_A 1}$ represents wave-function of electron 1 on atom H_A and $\psi_{H_B 2}$ represents wave-function of electron 2 on atom H_B.

and the total energy of the system is equal to the sum of energies of the separate atoms.

$$E_{total} = E_{H_A 1} + E_{H_B 2}$$

Case: If the atoms are close, it is not possible to distinguish whether electron 1 or electron 2 is on atom A (or atom B). When the two outcomes are equally probable, quantum mechanics instructs to describe the state of the system by linear combination of each possibility. So, the wave-function for the molecule is constructed by taking linear combination of both the possibilities as:

$$\psi_{VB} = \psi_{H_A 1}\psi_{H_B 2} + \psi_{H_A 2}\psi_{H_B 2} = c_1\psi_1 + c_2\psi_2$$

where $\psi_{H_A}\psi_{H_B}$ represents 1s orbitals (Atomic orbitals) of the two hydrogen atoms A and B respectively (the spins of the electrons are neglected for the sake of simplicity) and c_1 & c_2 represent the coefficients. The above wave-function represents covalent structures as each bonded atom contains one electron.

Another possibility of electron distribution that can be taken into account is when both the electrons belong either to atom A or atom. So, the wave-function representing ionic structures is given by the equation,

$$\psi_{Ionic} = \psi_{H_A 1}\psi_{H_A 2} + \psi_{H_B 1}\psi_{H_B 2}$$

with this total VB wave-function becomes,

$$\psi_{VB} = \psi_{Covalent} + \psi_{Ionic} = \psi_{H_A 1}\psi_{H_B 2} + \psi_{H_A 2}\psi_{H_B 1} + \psi_{H_A 1}\psi_{H_A 2} + \psi_{H_B 1}\psi_{H_B 2}$$

Energy Expression for Hydrogen Molecule

Taking into consideration the covalent structure, the energy value of hydrogen molecule is given by,

$$E_a = \frac{\int \psi_{VB}\,\widehat{H}_e\,\psi_{VB}\,d\tau}{\int \psi_{VB}\,\psi_{VB}\,d\tau}$$

VSEPR Theory

The Valence Shell Electron Pair Repulsion Theory abbreviated as VSEPR theory is based on the premise that there is a repulsion between the pairs of valence electrons in all atoms, and the atoms will always tend to arrange themselves in a manner in which this electron pair repulsion is minimalized. This arrangement of the atom determines the geometry of the resulting molecule.

The different geometries that molecules can assume in accordance with the VSEPR theory can be seen in the illustration provided below.

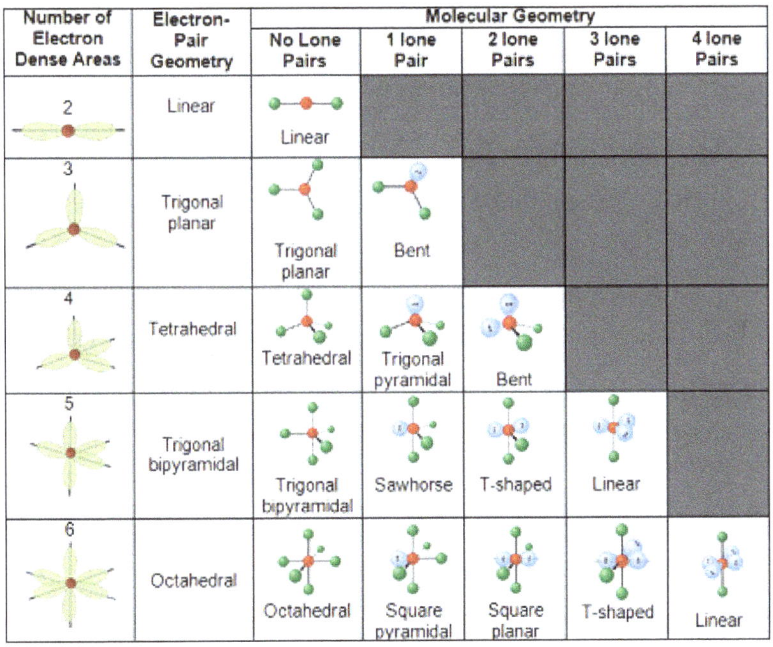

Number of Electron Dense Areas	Electron-Pair Geometry	Molecular Geometry				
		No Lone Pairs	1 lone Pair	2 lone Pairs	3 lone Pairs	4 lone Pairs
2	Linear	Linear				
3	Trigonal planar	Trigonal planar	Bent			
4	Tetrahedral	Tetrahedral	Trigonal pyramidal	Bent		
5	Trigonal bipyramidal	Trigonal bipyramidal	Sawhorse	T-shaped	Linear	
6	Octahedral	Octahedral	Square pyramidal	Square planar	T-shaped	Linear

VSEPR Theory – Different Geometries that Molecules can Assume.

The two primary founders of the VSEPR theory are Ronald Nyholm and Ronald Gillespie. This theory is also known as the Gillespie-Nyholm theory to honour these chemists.

According to the VSEPR theory, the repulsion between two electrons is caused by the Pauli exclusion principle that has greater importance than electrostatic repulsion in the determination of molecular geometry.

Postulates of VSEPR Theory

The postulates of the VSEPR theory are listed below:

- In polyatomic molecules (i.e. molecules made up of three or more atoms), one of the constituent atoms is identified as the central atom to which all other atoms belonging to the molecule are linked.
- The total number of valence shell electron pairs decides the shape of the molecule.
- The electron pairs have a tendency to orient themselves in a way that minimizes the electron-electron repulsion between them and maximizes the distance between them.

- The valence shell can be thought of as a sphere wherein the electron pairs are localized on the surface in such a way that the distance between them is maximized.
- Should the central atom of the molecule be surrounded by bond pairs of electrons, then, the asymmetrically shaped molecule can be expected.
- Should the central atom be surrounded by both lone pairs and bond pairs of electrons, the molecule would tend to have a distorted shape.
- The VSEPR theory can be applied to each resonance structure of a molecule.
- The strength of the repulsion is strongest in two lone pairs and weakest in two bond pairs.
- If electron pairs around the central atom are closer to each other, they will repel each other. This results in an increase in the energy of the molecules.
- If the electron pairs lie far from each other, the repulsions between them will be less and eventually, the energy of the molecule will be low.

Limitations of VSEPR Theory

Some significant limitations of the VSEPR theory include:

- This theory fails to explain isoelectronic species (i.e. elements having the same number of electrons). The species may vary in shapes despite having the same number of electrons.
- The VSEPR theory does not shed any light on the compounds of transition metals. The structure of several such compounds cannot be correctly described by this theory. This is because the VSEPR theory does not take into account the associated sizes of the substituent groups and the lone pairs that are inactive.
- Another limitation of VSEPR theory is that it predicts that halides of group 2 elements will have a linear structure, whereas their actual structure is a bent one.

Predicting the Shapes of Molecules

The following steps must be followed in order to decide the shape of a molecule:

- The least electronegative atom must be selected as the central atom (since this atom has the highest ability to share its electrons with the other atoms belonging to the molecule).
- The total number of electrons belonging to the outermost shell of the central atom must be counted.

- The total number of electrons belonging to other atoms and used in bonds with the central atom must be counted.
- These two values must be added in order to obtain the valence shell electron pair number or the VSEP number.

VSEP Number

The VSEP number describes the shape of the molecule, as described in the table provided below:

VSEP Number	Shape of the Molecule
2	Linear
3	Trigonal Planar
4	Tetrahedral
5	Trigonal Bipyramidal
6	Octahedral
7	Pentagonal Bipyramidal

However, the VSEPR theory cannot be used to obtain the exact bond angles between the atoms in a molecule.

Linear Shape of Molecule

- In this type of molecule, we find two places in the valence shell of the central atom.
- They should be arranged in such a manner such that repulsion can be minimized (pointing in the opposite direction).
- Example: BeF_2.

Trigonal Planar Shape of Molecule

- In this type of molecule, we find three molecules attached to a central atom.
- They are arranged in such a manner such that repulsion between the electrons can be minimized (toward the corners of an equilateral triangle).
- Example: BF_3.

Tetrahedral Shape of Molecule

- In two-dimensional molecules, atoms lie in the same plane and if we place these conditions on methane, we will get a square planar geometry in which the bond angle between H-C-H is 90°.

- Now, if we consider all these conditions for a three-dimensional molecule, we will get a tetrahedral molecule in which the bond angle between H-C-H is 109°28' (toward the corners of an equilateral triangle) CH_4.

Trigonal Bipyramid Shape of Molecule

- Let's take an example of PF_5. Here, repulsion can be minimized by even distribution of electrons towards the corner of a trigonal pyramid. In trigonal bipyramid, three positions lie along the equator of the molecule. The two positions lie along an axis perpendicular to the equatorial plane.

The strength of the repulsion between a lone pair and a bond pair of electrons lies in between the repulsion between two lone pairs and between two bond pairs. The order of repulsion between electron pairs is as follows:

Lone Pair- lone pair > Lone Pair- bond- pair > Bond Pair- bond pair.

- Total number of electron pairs around the central atom = ½ (number of valence electrons of central atom + number of atoms linked to central atom by single bonds):
    - For negative ions, add the number of electrons equal to the units of negative charge on the ions to the valence electrons of the central atom.
    - For positive ions, subtract the number of electrons equal to the units of positive charge on the ion from the valence electrons of the central atom.
- The number of Bond pair = Total number of atoms linked to central atom by single bonds.
- Number of lone pairs = Total number of electron – No of shared pair.

  The electron pairs around the central atom repel each another and move so far apart from each another that there are no greater repulsions between them. This results in the molecule having minimum energy and maximum stability.

    - The shape of a molecule with only two atoms is always linear.
    - For molecules with three or more atoms, one of the atoms is called the central atom and other atoms are attached to the central atom.
    - If the central atom is linked to similar atoms and is surrounded by bond pairs of electrons only, the repulsions between them are similar as a result the shape of the molecule is symmetrical and the molecule is said to have regular geometry.
    - If the central atom is linked to different atoms or is surrounded by bond pair

as well as a lone pair of electrons, the repulsion between them is similar. As a result, the shape of the molecule has an irregular or distorted geometry.

- The exact shape of the molecule depends upon the total number of electron pairs present around the central atom.

References

- Atomic-Theory, Supplemental-Modules-(Physical-and-Theoretical-Chemistry), Physical-and-Theoretical-Chemistry-Textbook-Maps: libretexts.org, Retrieved 28, June 2020
- Transition-state-theory, science: britannica.com, Retrieved 11, April 2020
- Introduction-to-the-transition-state-theory, introducing-the-effective-mass-of-activated-complex-and-the-discussion-on-the-wave-function-of-this-instanton: intechopen.com, Retrieved 25, July 2020
- Vsepr-theory: byjus.com, Retrieved 02, February 2020

3

Organic Chemistry

The branch of chemistry which deals with the study of organic compounds as well as their structure, reactions and properties is known as organic chemistry. The organic compounds are made up of a carbon atom bonded covalently with other atoms. The diverse reactions which are studied within organic chemistry such as electrophilic substitution reactions, SN1 reaction and SN2 reaction have been thoroughly discussed in this chapter.

Organic chemistry is the study of the structure, properties, composition, reactions, and preparation of carbon-contafining compounds, which include not only hydrocarbons but also compounds with any number of other elements, including hydrogen (most compounds contain at least one carbon–hydrogen bond), nitrogen, oxygen, halogens, phosphorus, silicon, and sulfur. This branch of chemistry was originally limited to compounds produced by living organisms but has been broadened to include human-made substances such as plastics. The range of application of Classification of Organic Compounds is enormous and also includes, but is not limited to, pharmaceuticals, petrochemicals, food, explosives, paints, and cosmetics.

Organic Compounds

Organic compound is the any of a large class of chemical compounds in which one or more atoms of carbon are covalently linked to atoms of other elements, most commonly hydrogen, oxygen, or nitrogen. The few carbon-containing compounds not classified as organic include carbides, carbonates, and cyanides.

Importance of Organic Compounds

- Organic compounds are important because all living organisms contain carbon.
- While carbohydrates, proteins and fats, the basic structures of life, are organic compounds.
- They are the basic components of many of the cycles that drive the earth. For example, the carbon cycle that includes the exchange of carbon between plants and animals in photosynthesis and cellular respiration.
- Organic compounds combine with metals to form organometallic compounds. These compounds are important industrially. They are used as catalysts, promoters, analysers as well as stabilizers.

Organic Reactions

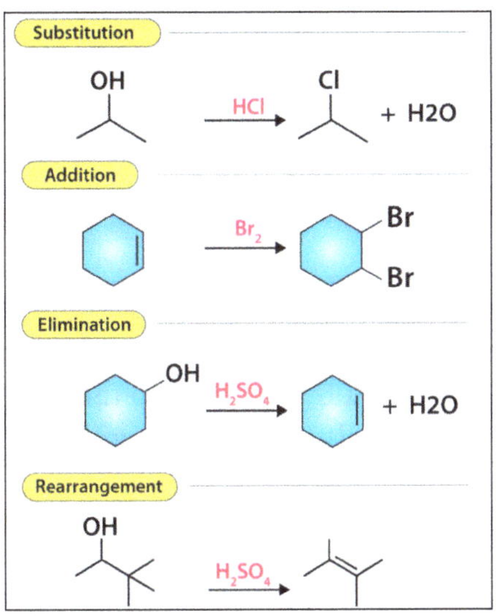

Organic reactions are the chemical reactions that are undergone by organic compounds (the chemical compounds containing carbon). A few important types of organic reactions are illustrated. There are mainly five types of organic reactions:

- Substitution reaction.
- Elimination reaction.
- Addition reaction.
- Radical reactions.
- Oxidation-Reduction Reactions.

Substitution Reactions

In a substitution reaction, one atom or a group of atoms is substituted by another atom or a group of atoms to form a new substance.

Let us take an example of a C-Cl bond, in which the carbon atom has a partial positive charge due to the presence of highly electronegative chlorine atom. In a nucleophilic substitution reaction, the nucleophile must have a pair of electrons and also should have a high affinity for the electropositive species as compared to the substituent which was originally present.

Elimination Reaction

There are some reactions which involve the elimination or removal of the adjacent atoms. After these multiple bonds are formed and there is a release of small molecules as products. One of the examples of elimination reaction is the conversion of ethyl chloride to ethylene.

$$CH_3CH_2Cl \rightarrow CH_2 = CH_2 + HCl$$

In the above reaction, the eliminated molecule is HCl, which is formed by the combination of H^+ from the carbon atom which is on the left side and Cl– from the carbon atom which is on the right side.

Addition Reactions

Addition reaction is nothing but just the opposite of elimination reaction. In an addition reaction, the components A and B are added to the carbon-carbon multiple bonds and this is called addition reaction. In the reaction given below when HCl is added to ethylene, it will give us ethylene chloride.

$$HCl + CH_2 = CH_2 \rightarrow CH_3CH_2Cl$$

Radical Reactions

Many of the organic reactions involve radicals. Addition of a halogen to a saturated hydrocarbon involves free radical mechanism. There are three stages involved in a radical reaction i.e. initiation, propagation, and termination. Initially when the weak bond is broken initiation of the reaction takes place with the formation of free radicals. After that when the halogen is added to the hydrocarbon a radical is produced and finally, it gives alkyl halide.

Oxidation and Reduction Reactions

Organic reactions are classified as oxidation or reduction depending on whether the substrate is oxidised or reduced. For this the functional groups are arranged in order of increasing oxidation state as follows:

Oxidation state -4 (alkane)

Oxidation state -2 (alkene, alcohol, alkyl halide, amine)

Oxidation state 0 (alkyne, carbonyl, dihalides, diols)

Oxidation state +2 (carboxylic acid, amide, trihalide)

Oxidation state +4 (carbon dioxide, carbon tetrachloride)

Oxidation is the conversion of a functional group in a molecule from one category to a higher one and reduction is the opposite. In oxidation gain of oxygen or/and loss of hydrogen takes place. Reduction is opposite to that of oxidation as here loss of oxygen or gain of hydrogen takes place. Also there is no oxidation without a concurrent reduction.

Oxidation reactions are classified depending on the type of bond change involved. Examples of oxidation reactions:

Elimination of Hydrogen

- Aromatisation: Conversion of cyclohexane to benzene in the presence of a catalyst is a type of oxidation reaction which proceeds via loss of hydrogen. The most frequently used reagents to affect aromatization are hydrogenation catalysts such as Pt, Pd, Ni etc. quinines, elements S, Se etc.

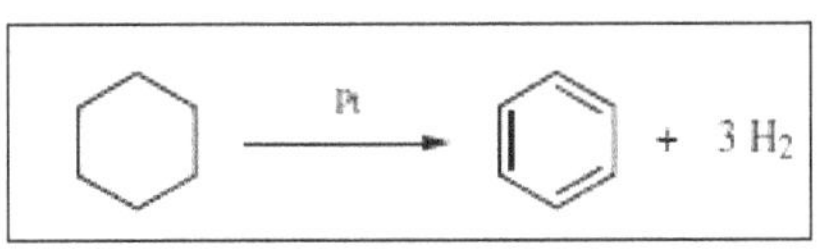

- Oxidation or dehydrogenation of alcohols to aldehydes and ketones. Conversion of alcohol to carbonyl: A variety of reagents such as CrO3, pyridinium chlorochromate (PCC), pyridinium dichromate (PDC), DMSO/oxalyl chloride, DMSO/DCC, copper chromite, are used for oxidation of alcohols to aldehyde or ketone. Primary alcohols are converted to aldehydes and secondary alcohols are converted to ketones in four main ways:
    - With strong oxidizing agents.
    - By catalytic dehydrogenation.
    - With NBS and other chemoselective oxidising agents.

$$RCH_2OH \xrightarrow[\text{chromite}]{\text{copper}} RCHO$$

$$RCHOHR' \xrightarrow[H_2SO_4]{K_2Cr_2O_7} RCOR'$$

- Cleavage of carbon - carbon bonds i.e. Ozonolysis: Reaction of alkenes with ozone under suitable conditions leads to fission of carbon-carbon double bond via formation of an intermediate ozonide.

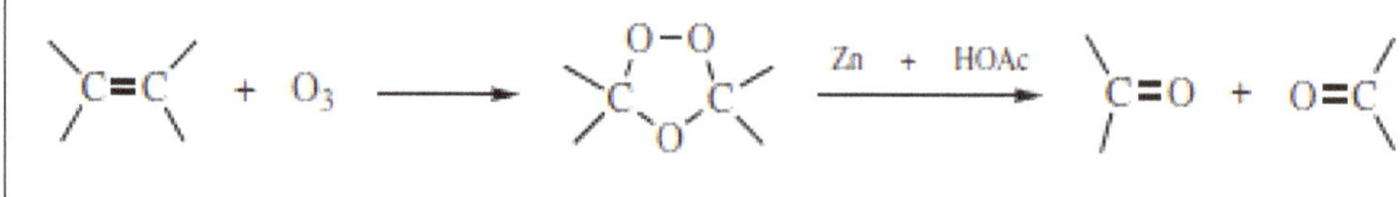

16

Examples of Reduction Reaction

- Conversion of carbonyl to alcohol: When aldehydes or ketones are converted to alcohol the hydrogen content is increased thus it is a reduction reaction. The transfer of hydride ion by lithium hydride to carbonyl is such an example. There are two methods known for the conversion of carbonyl to alcohol, these are Clemmensen reduction and the wolff kishner reduction.

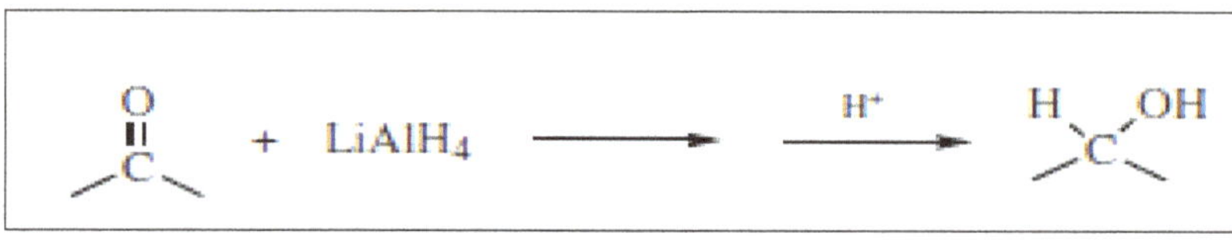

- Meerwein–Ponndorf–Verley reduction: It occurs in the presence of isopropyl alcohol and aluminum isopropoxide. The reaction is reversible in nature and the reaction reverse to it is named as the Oppenauer oxidation.

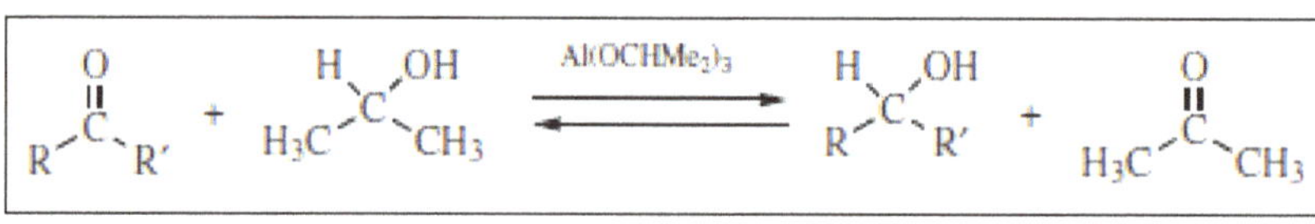

- Clemensen Reduction: In the Clemensen reduction a carbonyl compound is reduced to alkane using amalgamated zinc and hydrochloric acid. Ketones are reduced more often than aldehydes.

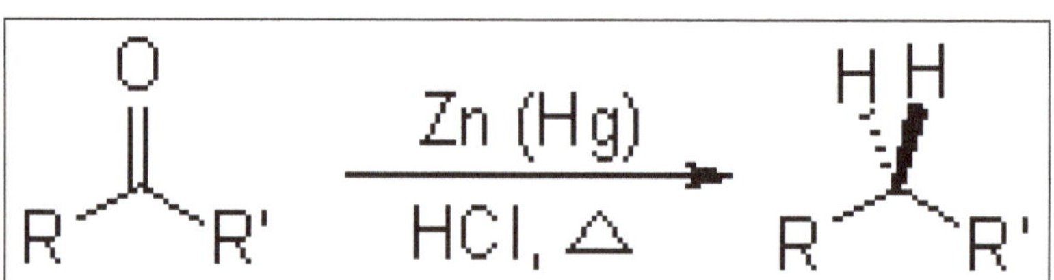

- Conversion of carboxylic ester to alcohol: Lithium hydride often brings out reduction of carboxylic ester by hydride transfer via formation of aldehyde in first step and further hydride transfer to the aldehyde leads to formation of alcohol. This reaction is of wide scope and has been used to reduce many esters. Other common reagents used are DIBAL, Lithium triethylborohydride, BH_3-SMe_2 in refluxing THF. $NaBH_4$ reduces phenolic esters especially those containing EWGs.

$$RCOOR' \xrightarrow{LiA1H_4} RCH_2OH + R'OH$$

- Conversion of nitro compounds to amine: Conversion of nitro group to amine is a reduction reaction which involves loss of oxygen and gain of hydrogen.

$$RNO_2 \xrightarrow[HCl]{Zn} RNH_2$$

Both aliphatic and aromatic nitro compounds can be reduced to amines, though the reaction has been applied much more often to aromatic nitro compounds. The other common reagents are Zn, Sn or/and acid, AlH_3-$AlCl_3$, $TiCl_3$, Al-$NiCl_2$-THF, formic acid and Pd-C, etc.

S_N1 Reaction

Nucleophilic substitution reactions that follow first order kinetics are known as S_N1 reactions. The rate of first order reaction depends only upon the concentration of the substrate. These reactions can be completed in following manner:

- C-X bond undergoes heterolysis in the first step (dissociation).
- The positively charged carbon that is produced in the first step is called a carbocation and is of major importance in S_N1 reactions.
- In the second step, nucleophile (Nu) forms bonds to the carbon of carbocation (heterogenesis).

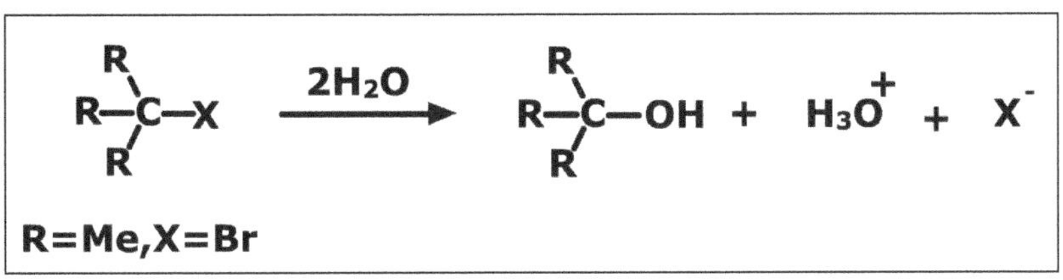

The mechanism of these reactions as follows:

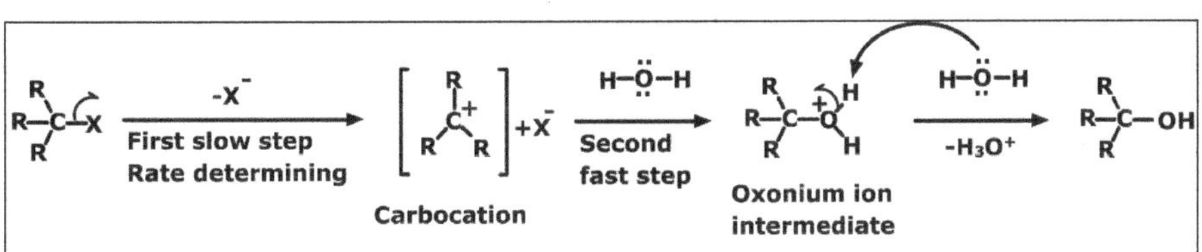

Kinetics of S_N1 Reactions

The S_N1 reaction is initiated by the dissociation of the leaving group and formation of the carbocation intermediate in the first step. After formation of the carbocation intermediate, the nucleophile takes part in the second step. Increasing or decreasing the concentration of the nucleophile has no measurable effect on the rate. The nucleophile is not involved in the initial step of rate-determination, thus the concentration does not affect the overall reaction rate. This reaction follows first order kinetics. The reaction is first order with respect to alkyl halide while zero order with respect to the nucleophile.

Rate = k [Substrate]

The energy profile diagram is shown in figure.

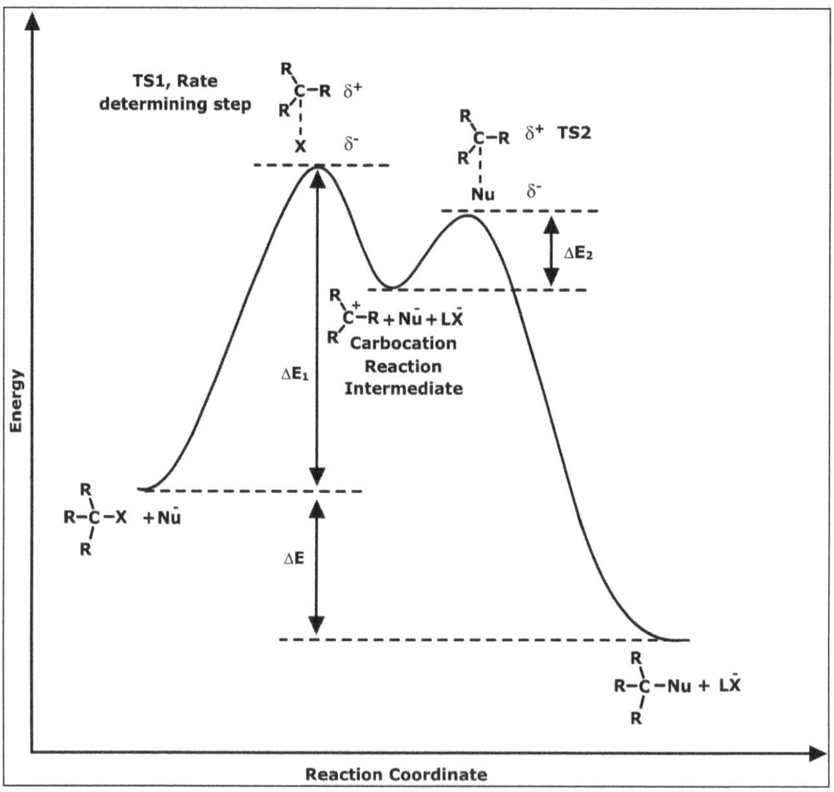

The S_N1 reactions are sometimes referred to solvolysis reactions.

Stereochemical Features of S_N1 Reaction

An enantiomerically pure product, (that is, one enantiomer), these reactions tend to result in a mixture of products where the stereochemistry is the same as the starting material (retention) or opposite (inversion). In other words, some degree of racemization will take place.

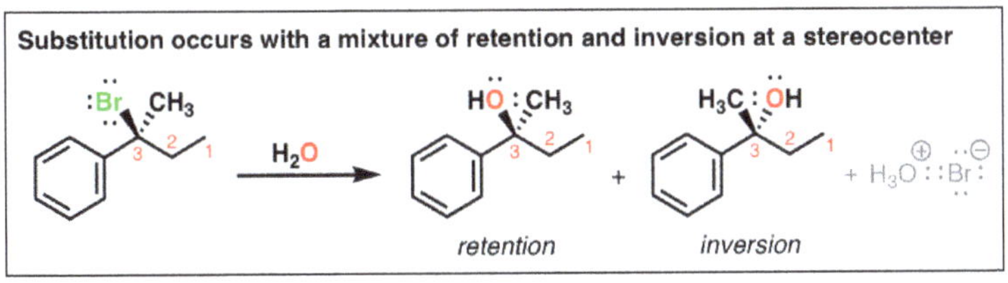

The Rate Law of the S_N1 Reaction Is First-Order Overall

We can also measure the rate law of these reactions. When we do so, we notice that the rate is only dependent on the concentration of the substrate, but not on the concentration of nucleophile.

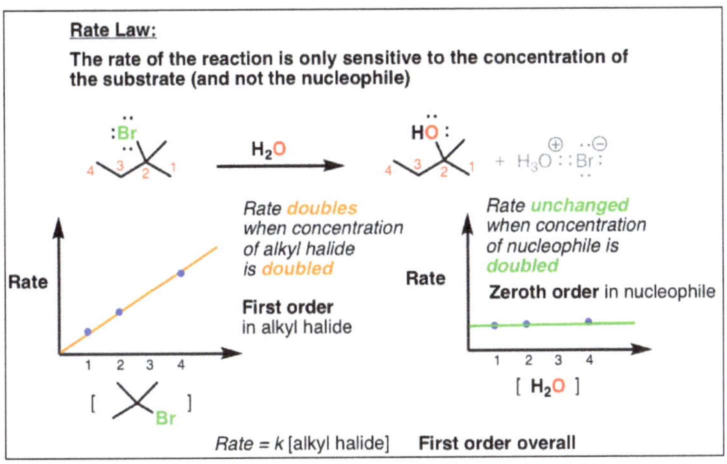

The Reaction Rate Increases with Substitution at Carbon

When we subtly change the types of substrates (e.g. alkyl halides) we use in these reactions, we find that tertiary substrates (for instance, *t*-butyl bromide) are considerably **faster** than secondary alkyl bromides, which are in turn faster than primary.

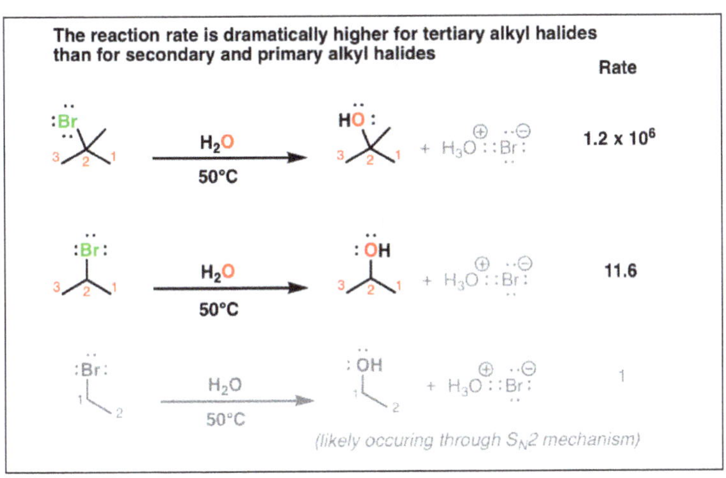

Compare that to the case for S_N2, where primary was faster than secondary and tertiary hardly reacted at all.

The Stepwise Reaction Mechanism of the S_N1 Reaction

The best hypothesis we have for this reaction is the *stepwise mechanism*. In the first step, the leaving group leaves, forming a carbocation. In the second, a nucleophile attacks the carbocation, forming the product.

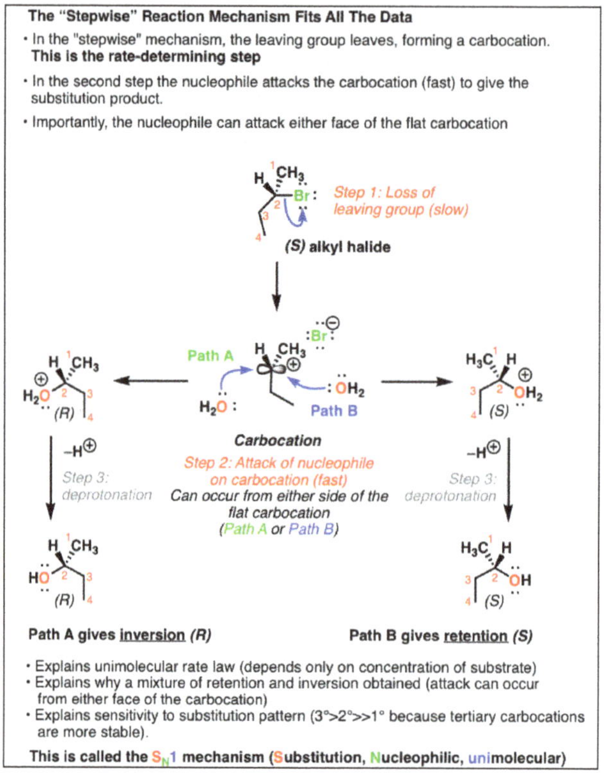

This explains all of our observations nicely. First of all, the slow step should be formation of the (unstable) carbocation – which only depends on the substrate, not the nucleophile. Furthermore, since the stability of carbocations depends tremendously on substitution pattern (tertiary carbocations are more stable than secondary, which are more stable than primary) this also conveniently explains the dependence of the reaction rate on substitution pattern. Any factor which stabilizes the carbocation, increases the rate at which the leaving group can leave.

Factors Influencing S_N1 Reaction

Solvent

The rate of the reaction can be affected by the energy level of the reagents. Solvation of the carbocation allows the carbocation to be surrounded by more electron density, making the positive charge more stable. The solvent can be protic or aprotic, but it must be polar solvent. Polar protic solvents have a H-atom attached to an electronegative atom so the hydrogen is highly polarized. Polar aprotic solvents have a dipole moment, but their hydrogen is not highly polarized. Polar aprotic solvents are not used in S_N1 reactions because some of them can react with the carbocation intermediate and lead to unwanted product. Thus, polar protic solvents are preferred in S_N1 reaction which helps to speed up the rate of reaction due to large dipole moment of the solvent which helps to stabilize the transition state. The highly positive and highly negative parts interact with the substrate to lower the energy of the transition state. Since the carbocation is unstable, anything that can stabilize this even a little will speed up the reaction.

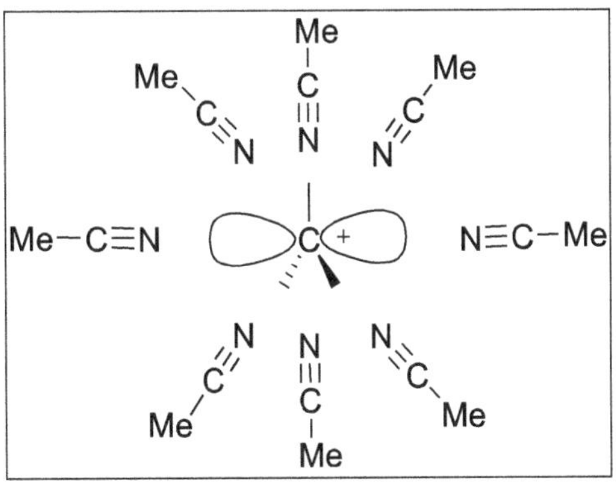

Sometimes in an S_N1 reaction the solvent acts as the nucleophile known as solvolysis reaction. The polarity and the ability of the solvent to stabilize the intermediate carbocation, is very important as shown by the relative rate data for the solvolysis (in table). The dielectric constant of a solvent provides a measure of the solvent's polarity. A dielectric constant below 15 is usually considered non-polar. Thus, higher the dielectric constant more polar will be substance and in the case of S_N1 reactions the faster the rate.

Solvent	Dielectric Constant	Relative rate
Acetic acid	6	1
Methanol	33	4
Water	78	150000

The example given here illustrates Solvent.

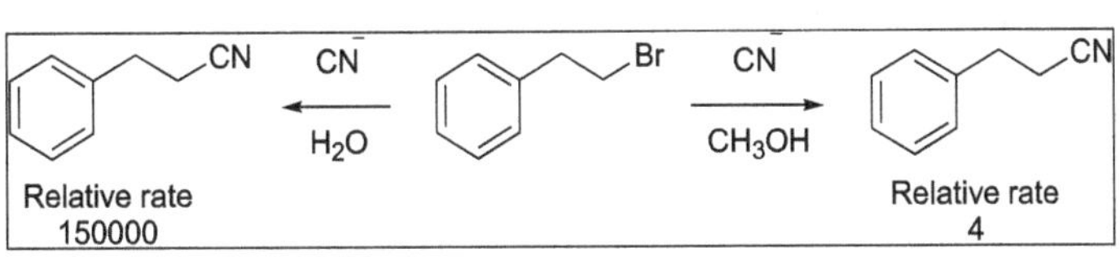

Nucleophile

Nucleophiles involved in the S_N1 mechanism are mostly weak and neutral molecules (viz. H_2O, ROH). The strength of the nucleophile does not affect the reaction rate of S_N1

because; the nucleophile is involved in the rate-determining step. However, if you have more than one nucleophile competing to bond to the carbocation, the strengths and concentrations of those nucleophiles affect the distribution of products.

For example, if you have $(CH_3)_3CCl$ reacting in water and formic acid where the water and formic acid are competing nucleophiles, you will get two different products: $(CH_3)_3COH$ and $(CH_3)_3COCOH$. The relative yields of these products depend on the concentrations and relative reactivities of the nucleophiles.

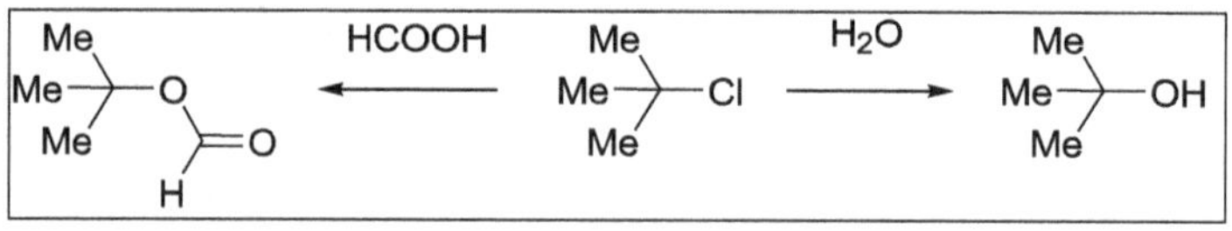

Substrate

The hyperconjugation and the inductive effect allow alkyl groups to stabilize carbocations. The more stable carbocation intermediate has a lower activation barrier, so the S_N1 reaction occurs faster. In general, the S_N1 reaction is favored in the order is,

Benzylic > Allylic > 3° > 2° > 1° >> Me^+.

Substrates wherein the leaving group (LG) is on a 3° carbon will lead to a reaction in the presence of a good nucleophile (Nu). Since, 1° carbocations are highly energetically unfavorable, as a rule of thumb they generally do not form. The rate for hydrolysis of alkyl halide is in the order of halides, 3o > 2° > 1° > MeX. Thus, we can say the electronic factor is more important than the steric factor. The Eact for the carbocation intermediate will be highest for MeX (1°), while least for the 3°. The molecules in which carbon next to the site of substitution contains a double bond, the SN1 reaction is possible. The reason is that the positive charge on the carbocation can be delocalized among multiple possible resonance structures (resonance and delocalization) making the carbocation dramatically stable. This effect can occur when the carbon atom of interest is next to one double bond (allylic) or a benzene ring (benzylic). In allylic case the delocalization of the positive charge, the nucleophile can attack at multiple sites while this effect is absent in the benzylic system due to the need to preserve aromaticity. For example, the allylic carbocation can form two different resonance structure, both are available for reaction. In the first example we end up with similar carbocation intermediate but we have different situation in the second example where we have 2° and 3° carbocation intermediates. Thus, second reaction will lead to 3o product through stable 3° carbocation intermediate.

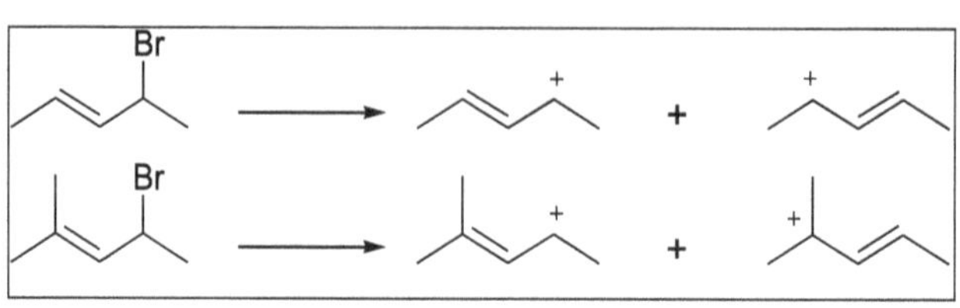

Delocalization and resonance.

If there is a benzylic carbocation, it is also resonance stabilized but only the carbocation on the benzylic position is reactive (retains the aromatic ring) as follows:

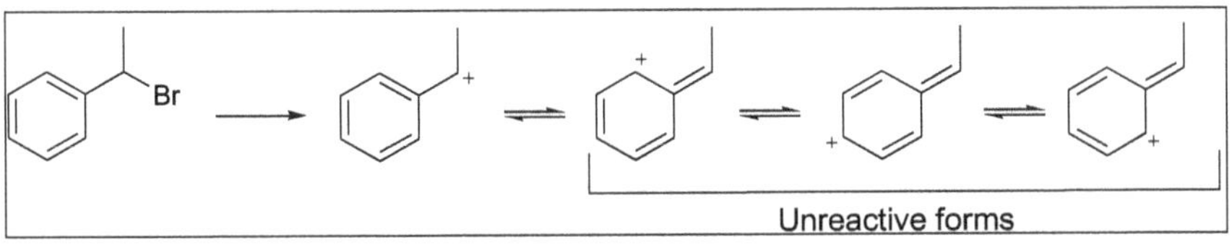

Leaving Group

The leaving group is almost always expelled with a full negative charge. The best leaving groups are those that can best stabilize an anion (i.e. a weak base). S_N1 reaction speeds up with a good leaving group. This is because the leaving group is involved in the rate-determining step. A good leaving group wants to leave so it breaks the C-leaving group bond faster. Examples of LG:

(Good.......) $^-$OMs, $^-$OTos, Triflate ion, NH_3 > H_2O ≈ I^-, Br^- > Cl^- > F^- ($^-$OH, $^-NH_2$) (.......poor)

As you go from left to right on the periodic table, electron donating ability decreases and thus ability to be a good leaving group increases. Halides are an example of a good leaving group whose leaving-group ability increases as you go down the column. Other examples of good leaving group viz. methyl sulfate ion and other sulfonate ions.

(Good)..... I^- > Br^- > Cl^- > F^- .....(poor)

For example: The two reactions below is the same reaction done with two different leaving groups. One is significantly faster than the other. This is because the better leaving group leaves faster and thus the reaction can proceed faster.

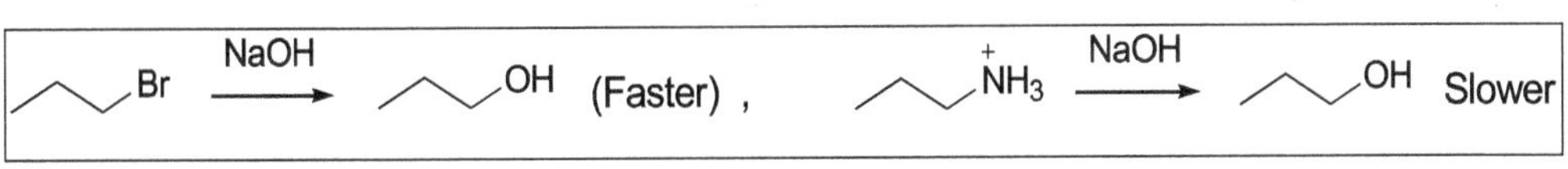

Examples of S_N1 Reactions

Bromination: S_N1 Mechanism for Reaction of Alcohols with HBr

Step 1: Acid/base Reaction: Protonation of the alcoholic oxygen to make it a better leaving group. This step is fast and reversible. The lone pairs on the oxygen make it a Lewis base.

Step 2: Rate determining step: Cleavage of the C-O bond allows the loss of the leaving group, a neutral water molecule, to give a carbocation intermediate.

Step 3: Attack of the nucleophilic bromide ion on the electrophilic carbocation creates the alkyl bromide.

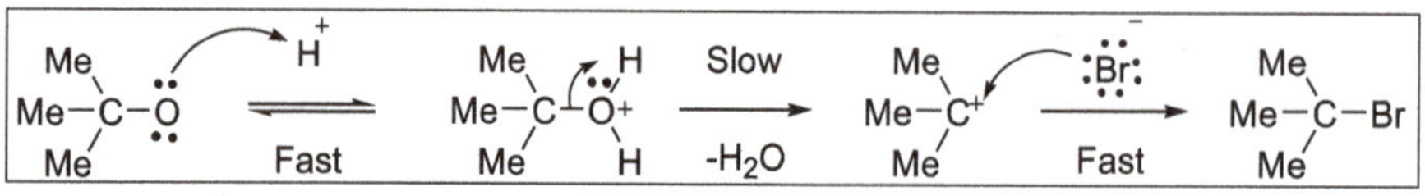

Rearrangement Reactions in S_N1 Reactions

Carbocation intermediates, once formed rearrangements can potentially occur which include S_N1 reactions, E1 elimination and electrophilic addition to double bonds. Some examples are as follows which involve either hydride shift or alkyl shift.

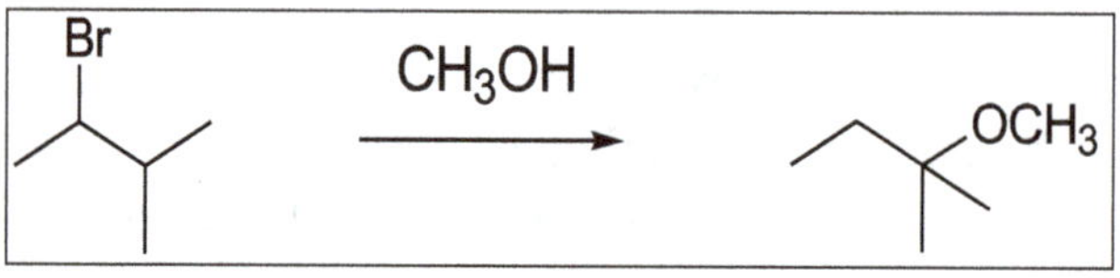

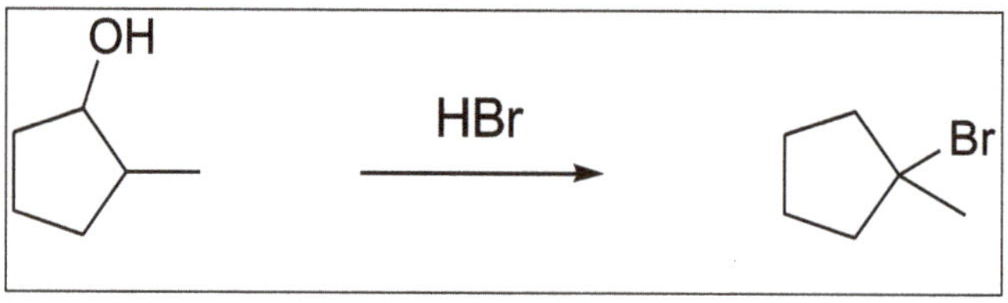

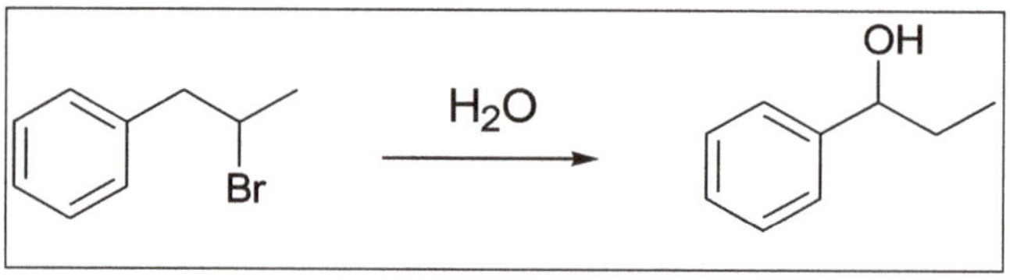

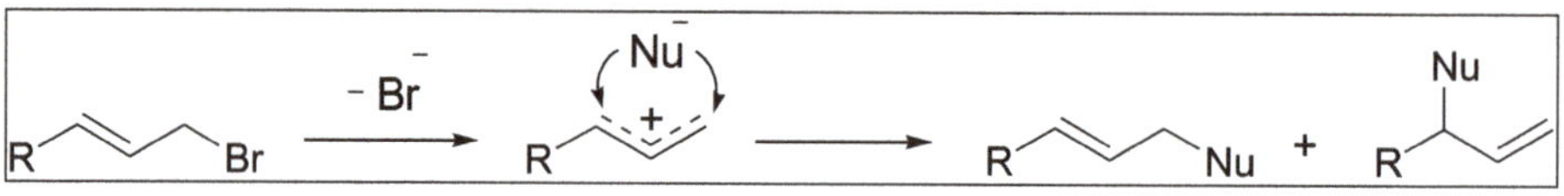

S_N2 Reaction

S_N2 indicates a substitution, nucleophilic, bimolecular reaction, described by the expression rate = k [Nu][R-LG]. This implies that the rate determining step involves an interaction between two species, the nucleophile and the organic substrate. This pathway is a concerted process (single step) as shown by the following reaction coordinate diagrams, where there is simultaneous attack of the nucleophile and displacement of the leaving group.

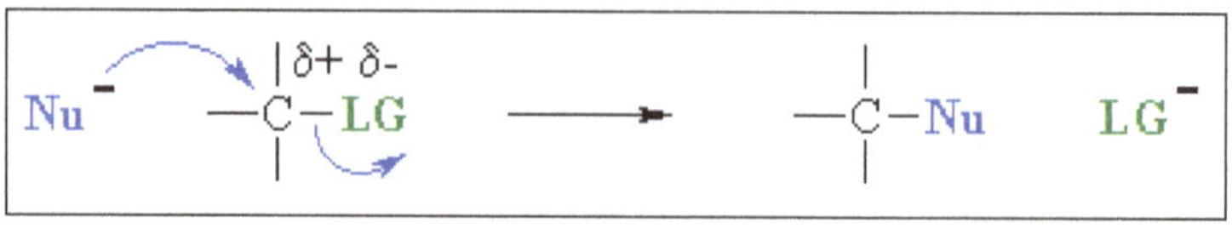

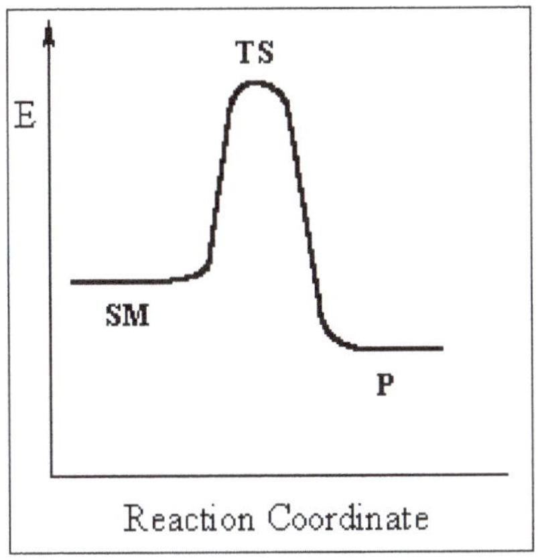

The nucleophile attacks at the carbon with the partial positive charge as a result of the polar s bond to the electronegative atoms in the leaving group.

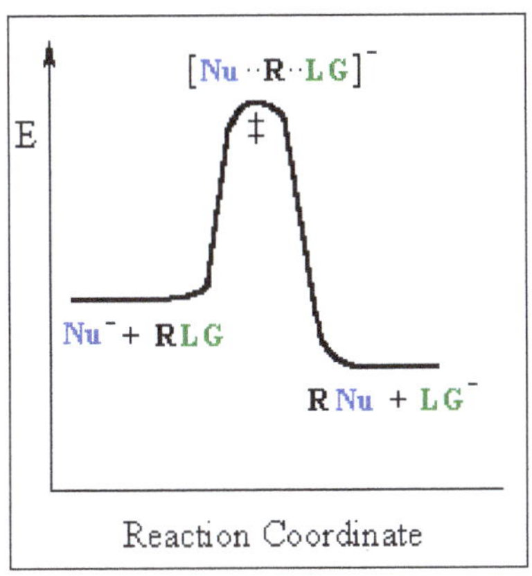

Single step reactions have no intermediates and a single transition state (TS). In an S_N2 there is simultaneous formation of the carbon-nucleophile bond and breaking of the carbon-leaving group bond, hence the reaction proceeds via a TS in which the central C is partially bonded to five groups.

General Case

R- Reactivity order: $CH_3^- > CH_3CH_2^- > (CH_3)_2CH^- > (CH_3)_3C^-$

In an S_N2 reaction, the transition state has 5 groups around the central C atom. As a consequence of the steric requirements at this center, less highly substituted systems (*i.e.* more smaller H groups) will favour an S_N2 reaction by making it easier to achieve the transition state.

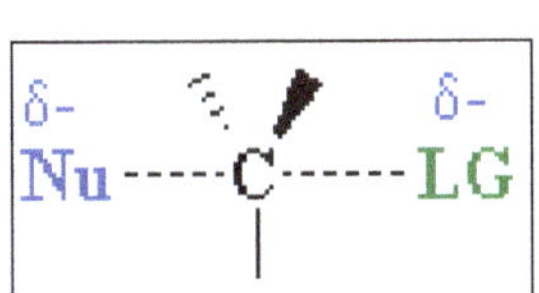

The following two series of JSMOL images show four alkyl bromides and the iodide ion as the nucleophile with relative rate of reaction data.

Use the lower row of space filling models (which are great for seeing steric effects) to rotate the molecules to look at the electrophilic C center from the side opposite to the leaving group (which is where the nucleophile attacks from) to see how much of the electrophilic center you can see. In order to do this, you need to rotate the model so that the Br atom (maroon) is as much out of site as possible. If you need help recognising the electrophilic center, use the button to highlight it.

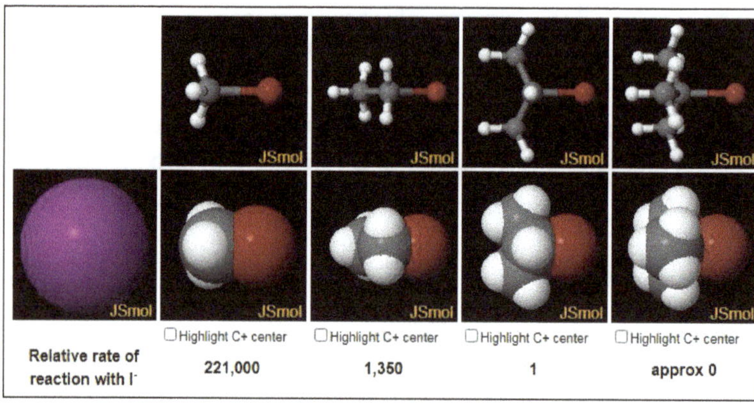

Notice how as the steric crowding increases around the electrophilic center that the rate of reaction with iodide decreases.

-LG

The C-LG bond is broken during the rate determining step so the rate does depend on the nature of the leaving group. However, if a leaving group is too good, then an S_N1 reaction may result.

Nu

Since the nucleophile is involved in the rate determining step, the nature of the nucleophile is very important in an S_N2 reaction. The more reactive the nucleophile, the more likely the reaction will be S_N2 rather than S_N1.

Stereochemistry

When the nucleophile attacks in an S_N2 it is on the opposite side to the position of the leaving group. As a result, the reaction will proceed with an inversion of configuration.

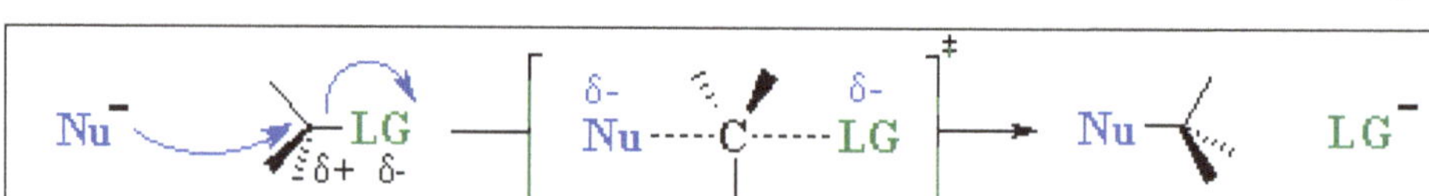

Solvent

Polar aprotic solvents can be used to enhance the reactivity of the nucleophile and help promote an S_N2 reaction.

S_Ni Reaction

The term S_Ni stands for substitution nucleophilic internal. In S_N1 and S_N2 reaction mechanism, the racemization and inversion of the configuration of the product take place. But Hughes, Ingold has shown that optically active 1-phenyl ethanol reacts with thionyl chloride to give 1-phenyl ethyl chloride with complete retention of configuration.

Normally the product should be inversion of configuration. so former can be explained by internal mechanism which is called S_Ni mechanism.

Example of S_Ni Reaction

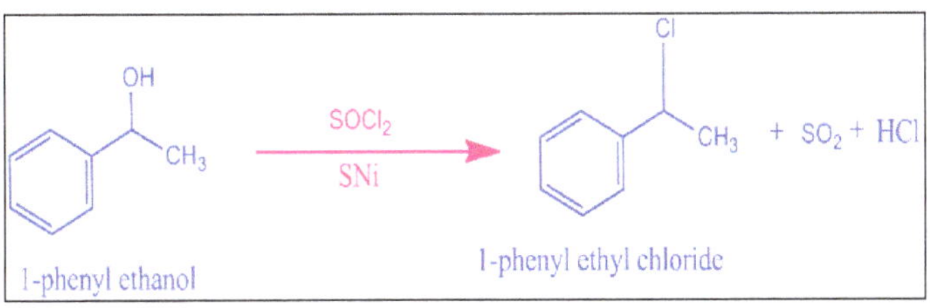

Mechanism of S_Ni Reaction

The solvents used in S_Ni reaction is diethyl ether or tetra hydro furan. The mechanism of SNi reaction shown below.

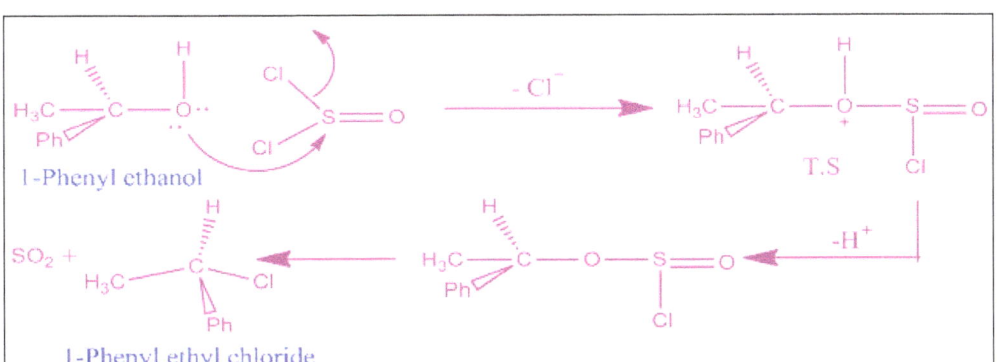

Stereochemistry of S_Ni Reaction

During the S_Ni reaction there occurs a internal transformation through a cyclic transition state to which attack of chloride from front side leads to retention in configuration. That is, both the reactant and products are having same configuration.

Reaction of 1-phenyl thanol with thionyl chloride is an example of S_Ni reaction. In this reaction,ether is used as a solvent. Pyridine does not used as a solvent instead of ether. Because, pyridine react with thionyl chloride to form pyridinium chloride. The chloride ion is free for attact on crabon atom of substrate.

Hence, chloride ion attact the carbon atom from back side. As a result, inversion of configuration of the product takes place. So the mechanism is not a S_Ni, the mechanism is actually SN_2.

Substitution Radical Nucleophilic Unimolecular

The $S_{RN}1$ (substitution radical nucleophilic unimolecular) reaction is defined by a unimolecular nucleophilic substitution reaction involving radical anions. The $S_{RN}1$ reaction was discovered by accident when a strange anomaly in the alkylation of aliphatic nitro compounds by alkyl halides was observed. In 1949 reseachers performed a routine alkylation of 2-nitropropane with 4- nitrobenzyl chloride. The product was mainly the C-alkylated compound A.

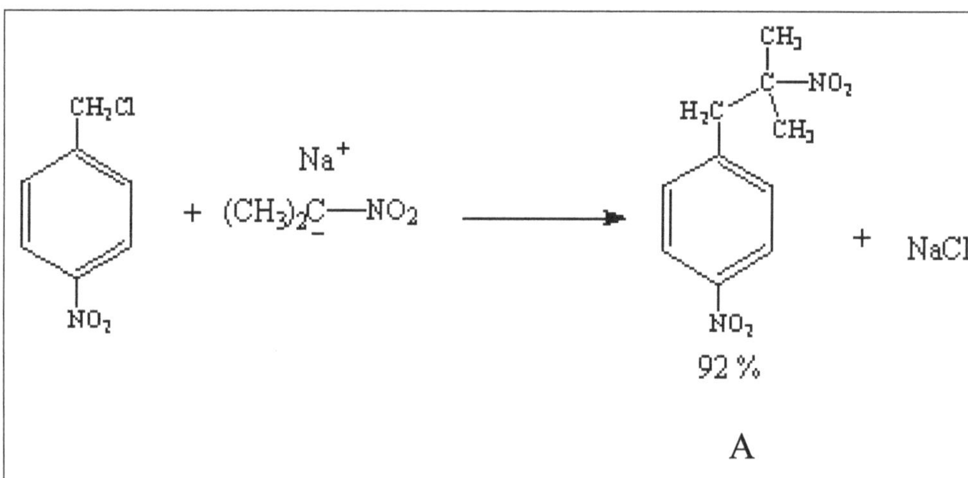

Reaction of 4-nitrobenzyl chloride with 2-nitropropane.

Due to involvement of radical anions, the SRN1 reactions may occur via a chain or non-chain mechanism. This reaction type was discovered in 1970 by Bunnett and Kim. The IUPAC designation for this kind of reactions is $T + D_N + A_N$.

It has been reported that the $S_{RN}1$ chain mechanism is initiated by single electron transfer (SET) from a chemical catalyst or an external electron source to the substrate which produces a radical anion. The general schematic for $S_{RN}1$ reactions is:

$$\text{initiation}: e^- + ArX \rightarrow [ArX]^{\bar{\cdot}}$$

$$\text{propagation}: \begin{cases} [ArX]^{\bar{\cdot}} \rightarrow Ar^{\cdot} + X^- \\ Ar^{\cdot} + Y^- \rightarrow [ArY]^{\bar{\cdot}} \\ [ArY]^{\bar{\cdot}} + ArX \rightarrow ArY + [ArX]^{\bar{\cdot}} \end{cases}$$

$$\text{overall}: ArX + Y^- \rightarrow ArY + X^-$$

As depicted above in figure, in the $S_{RN}1$ mechanism, the initial attack takes place via an electron donor, rather than a nucleophile. The radical anion thus formed leads to removal of leaving group and left behind a radical available for attack by the incoming nucleophile.

Specifically, for aromatic substrates, the reactions go by cine substitution although the need for electron withdrawing groups is not that stringent for these reactions. Interestingly, involvement of radicals allows a chain mechanism for $S_{RN}1$ reactions. A chain mechanism is a self-sustaining reaction, whereby, the reaction continues by the formation of reactive intermediates that recycle during product formation. These intermediates then react to continue additional processes keeping the continuity of reaction. Being a chain reaction, the mechanism follows an initiation, propagation and the termination steps.

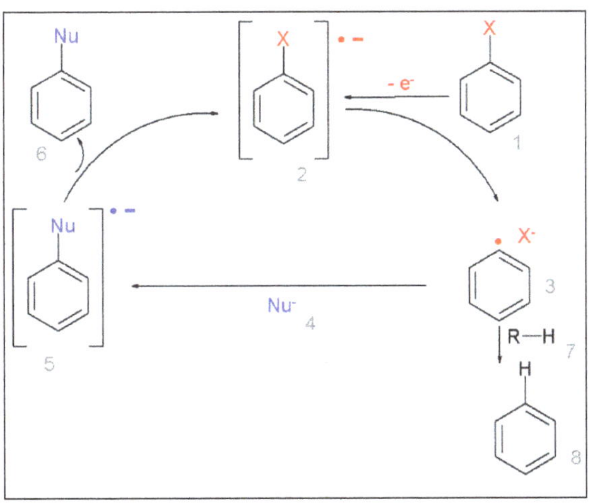

$S_{RN}1$ reaction with aryl halide as a substrate.

As shown in figure, the aryl halide substrate (1) first accept an electron from a donor and becomes an anion radical (2), which losses the halide anion giving rise to an aryl radical (3), which can be easily attacked by the nucleophile (4) to form a new radical anion (5), which subsequently leads to the product formation (6). During the chain propagation step, the aryl radical with attached nucleophile (5), get recycled to give intermediate radical (3) through intermediate (2). Alternatively, the phenyl radical (3) can also abstract a proton from an alternate RH species (7) to form the product (8) and leading to chain termination. The mechanism looks complicated and initially discovered through very interesting observations which were as follows:

When 5-iodo-1, 2, 4-trimethylbenzene was treated with a strong base such as KNH_2 in presence of NH_3, two products were formed with ratio of 0.63:1. By the time benzyne mechanisms were known and based on cine substitution, strong base and relatively less reactive substrate, this reaction was expected to go by this mechanism, however, benzyne mechanism would have lead to product formation in equal ratio which was not the case observed here.

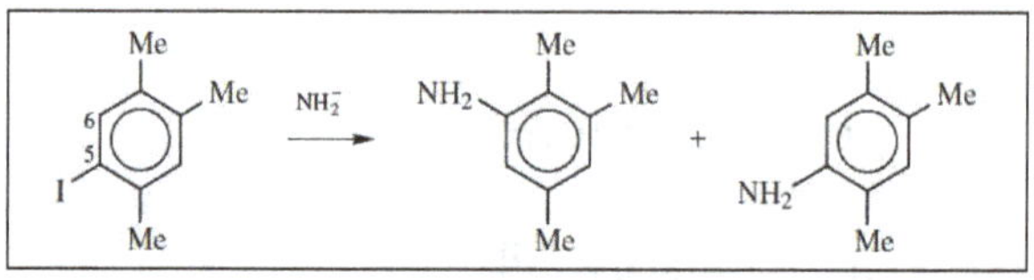

$S_{RN}1$ reaction in 5-iodo-1, 2, 4-trimethylbenzene.

Therefore, the reaction was suspected to go by a different mechanism. A radical mechanism was proposed for this reaction. The involvement of a radical mechanism was further confirmed by addition of radical scavenger that leads to different product ratios. To explain the iodo result, it has been proposed that in addition to the benzyne mechanism, a free radical mechanism is also operating here:

$$\mathrm{ArI} \xrightarrow[\text{donor}]{\text{electron}} \mathrm{ArI}^{\bullet-} \longrightarrow \mathrm{Ar}\bullet + \mathrm{I}^-$$

$$\mathrm{Ar}\bullet + \mathrm{NH_2}^- \longrightarrow \mathrm{ArNH_2}^{\bullet-} + \mathrm{ArI} \longrightarrow \mathrm{ArNH_2} + \mathrm{ArI}^{\bullet-}$$

followed by termination steps

Additionally, it was discovered that the $S_{RN}1$ reaction can occur via a non-chain mechanism as well. Non-chain $S_{RN}1$ reactions are less known. Non-chain $S_{RN}1$ reactions occur in solvent cages, which are defined as a region of space which includes the reacting species surrounded by solvent molecules. The solvent cage restricts the movement of reactive intermediates thereby not allowing a propagation, while, stepwise substitution occur giving rise to product formation only. Following is an example of non-chain $S_{RN}1$ reaction.

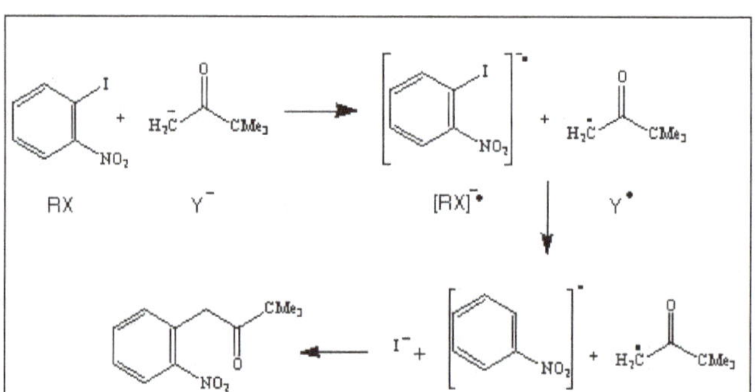

An example of non-chain $S_{RN}1$ reaction.

The substitution of iodo group with incoming reactive nucleophile occurs via the non-chain $S_{RN}1$ reaction as shown above.

A number of useful reactions proceed via this mechanism including photo stimulation reactions, electrochemical reactions and even some so no chemical reactions such as reaction of benzyl chloride and nitrobenzene. Initiation can also be achieved by a number of methods, the most commonly used being photo stimulation, introduction of solvated electrons or thermal reactions.

Reactions Involving $S_{RN}1$ Mechanism

As an example of photo stimulated reaction, the reactant, 3-chlorophenyltrimethylammonium iodide, upon reaction with Me_3SnNa (1) proceeded efficiently (30 min) to give 1, 3- bis(trimethylstannyl) benzene in good yield (85%).

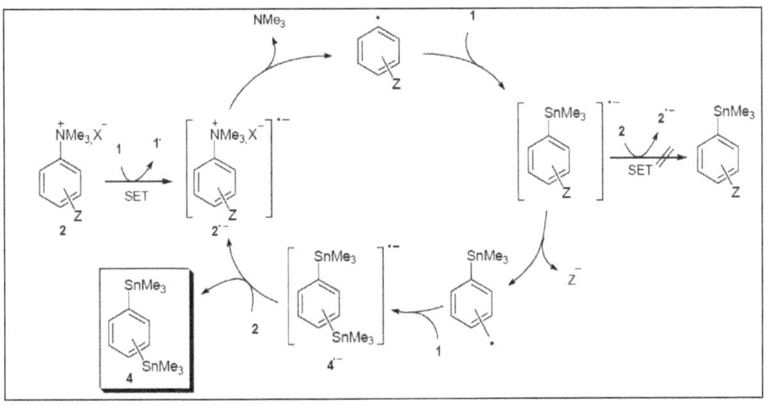

An example of photo stimulated reaction.

Interestingly, the reaction was slow in dark and completely quenched upon addition of pdinitrobenzene (p-DNB, 20%), a well-known inhibitor of radical anions. Thus, unambiguously establishing a $S_{RN}1$ mechanism, allowed by spontaneous electron transfer (ET) which was stimulated under irradiation.

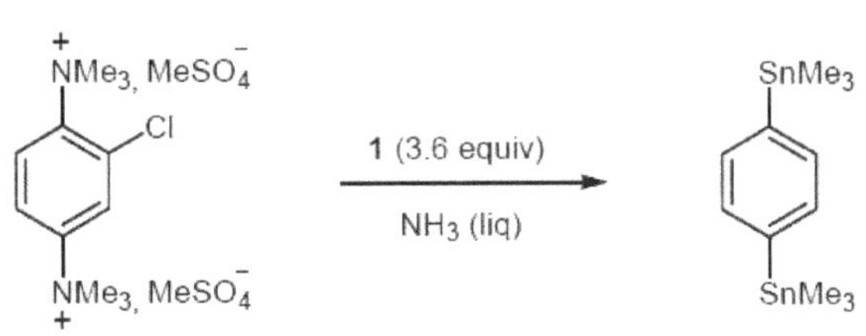

Conditions	Yield
hν, 30 min	85 %
hν, 5 min	55 %
dark, 30 min	10 %[a]
dark, 20 % *p*-DNB, 30 min	0 %[b]

Yields in photostimulated reaction.

A number of aromatic heterocyclic systems undergo reactions via SRN1 mechanisms. A few examples are shown below.

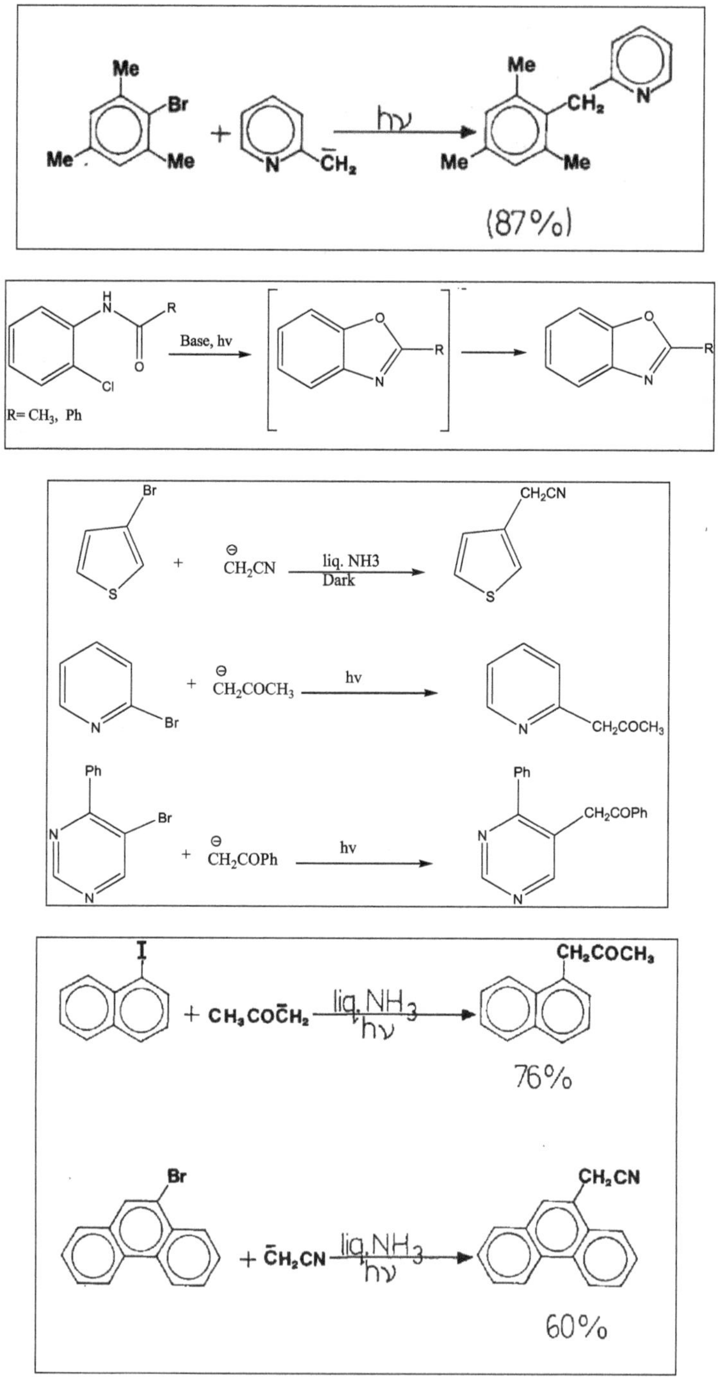

Examples of aromatic heterocyclic systems undergoing reactions via $S_{RN}1$ mechanisms.

The lithium salt of 2-nitropropane reacts with dinitro-cumene to afford an 87% yield of the substituted product within 3 hours.

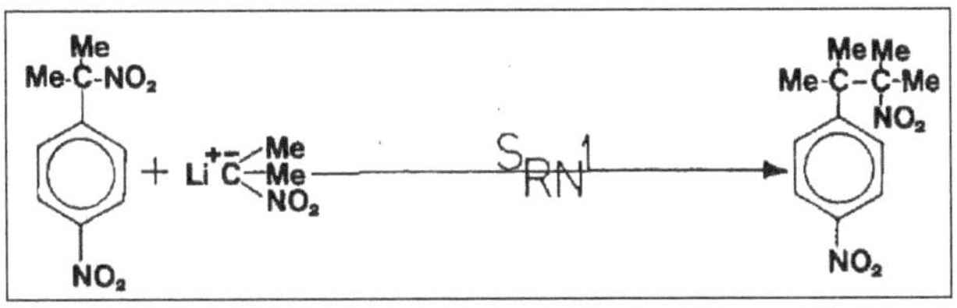

Reaction time with lithium salt of 2-nitropropane is only three hours.

The nucleophile-initiated substitution in dihaloarenes (shown below) goes very well by the variety of nucleophiles including ketone enolates, aromatic thiolates and dialkylphosphites.

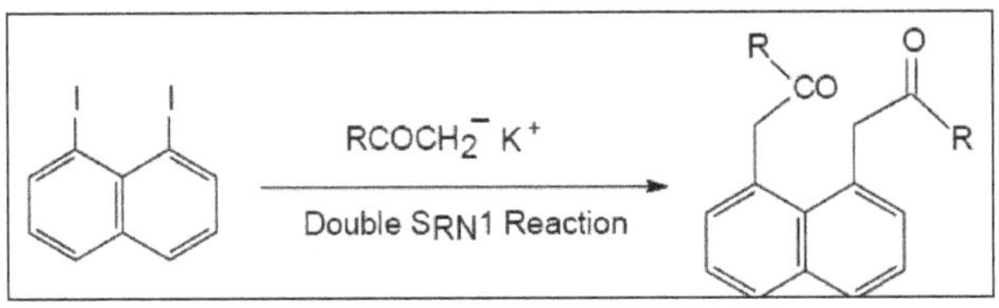

Nucleophile-initiated double $S_{RN}1$ substitution reaction.

Symmetric and asymmetric biaryls can also be synthesized by the electron transfer (ET) $S_{RN}1$ mechanism through the substitution of a haloaromatic substrate by β-naphthyl nucleophile such as the anion of 2-naphthol and 2-naphthylamine

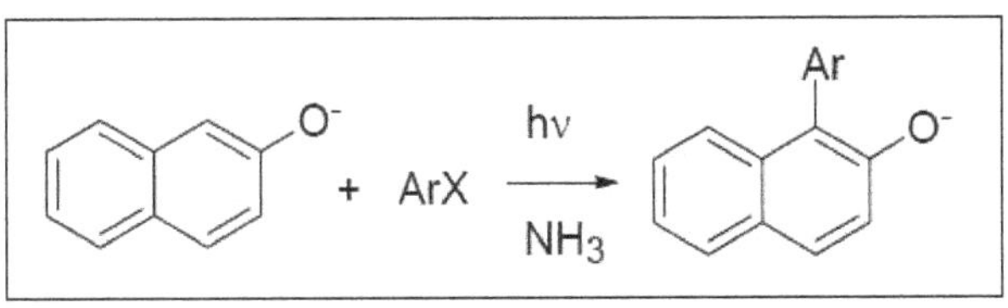

Symmetric and asymmetric biaryls synthesized by $S_{RN}1$ reaction.

The electrochemical reduction of aryl azosulfides yields the corresponding sulfides and decompositions products.

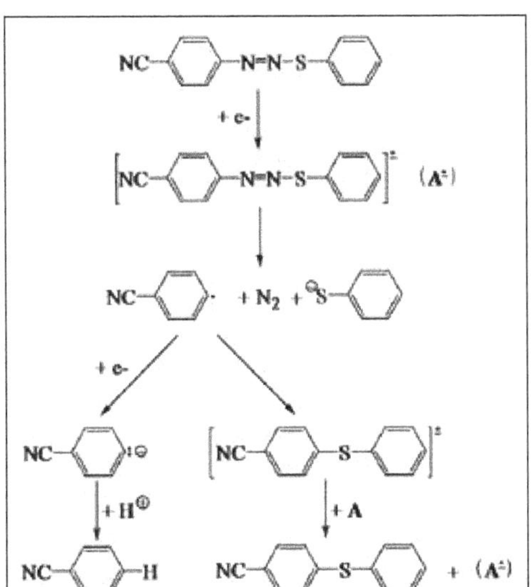

An example of chain $S_{RN}1$ mechanism reaction.

The reaction of 1, 2-dibromotetrafluoroethane with thiophenoxides in the presence of catalytic amount of SO_2 proceeds via $S_{RN}1$ mechanism. Specifically, the presence of SO_2 in the reaction promotes a $S_{RN}1$ process which results in quantitative yields of $ArSCF_2CF_2Br$ without the byproduct $ArSCF_2CF_2H$.

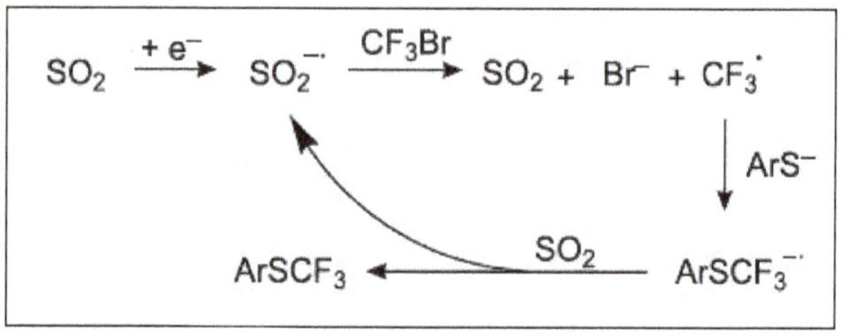

$S_{RN}1$ reaction in the presence of catalytic amount of SO_2.

Electrophilic Substitution Reactions

An electrophilic substitution reaction is a chemical reaction in which the functional group attached to a compound is replaced by an electrophile. The displaced functional group is typically a hydrogen atom. Electrophilic substitution reactions generally proceed via a three-step mechanism that involves the following steps:

- The generation of an electrophile.
- The formation of a carbocation (which is an intermediate).
- The removal of a proton from the intermediate.

Types of Electrophilic Substitution Reactions

The two primary types of electrophilic substitution reactions undergone by organic compounds are electrophilic aromatic substitution reactions and electrophilic aliphatic substitution reactions. An illustration describing the electrophilic substitution of a hydrogen atom (belonging to a benzene molecule) with a chlorine atom is provided below:

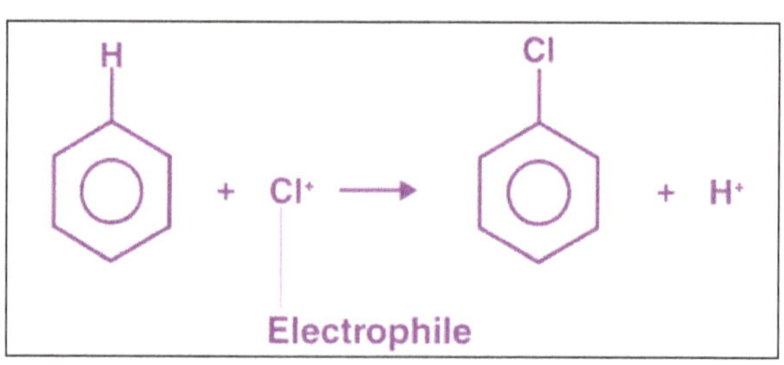

Here, the chlorine cation acts as an electrophile and replaces a hydrogen atom in the benzene ring. The products formed in this electrophilic substitution reaction include a proton and a chlorobenzene molecule.

Electrophilic Aromatic Substitution Reaction

In electrophilic aromatic substitution reactions, an atom attached to an aromatic ring is replaced with an electrophile. Examples of such reactions include aromatic nitrations, aromatic sulphonation, and Friedel-Crafts reactions.

It is important to note that the aromaticity of the aromatic compound is preserved in electrophilic aromatic substitutions. Therefore, these reactions can be used to obtain aryl halides from aromatic rings and iodine, bromine, or chlorine.

Electrophilic Aliphatic Substitution Reaction

In electrophilic aliphatic substitution reactions, an electrophile replaces the functional group (generally hydrogen) in an aliphatic compound. These reactions can be classified into the following five types.

- Halogenation of ketones.
- Nitrosation.
- Keto-Enol tautomerism.
- Insertion of a carbene into a carbon-hydrogen bond.
- Diazonium coupling (aliphatic).

These electrophilic substitution reactions can result in an inversion of configuration if the electrophilic attack occurs at an angle of 180° to the leaving group (attack from the rear).

Mechanism of Electrophilic Substitution Reaction

The electrophilic substitution reaction mechanism involves three steps.

Step 1: Generation of Electrophile

Anhydrous aluminium chloride is a very useful Lewis acid in the generation of electrophile from the chlorination, alkylation, and acylation of an aromatic ring. The resulting electrophiles (from the combination of anhydrous aluminium chloride and the attacking reagent) are Cl^+, R^+, and RC^+O respectively as shown below:

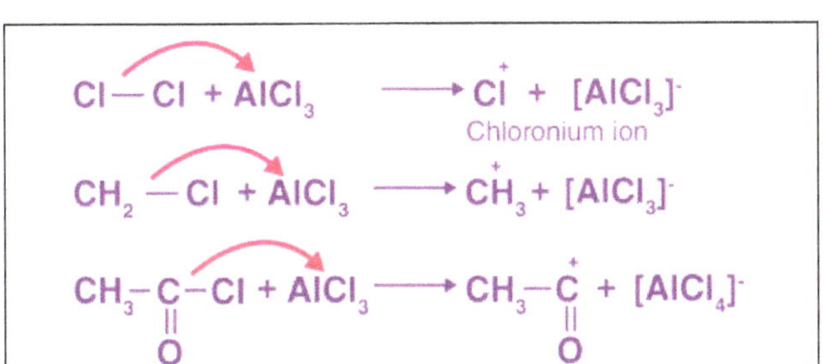

Step 2: Formation of Carbocation

The electrophile attacks the aromatic ring, forming a sigma complex or an arenium ion. One of the carbons in this arenium ion is sp^3 hybridized.

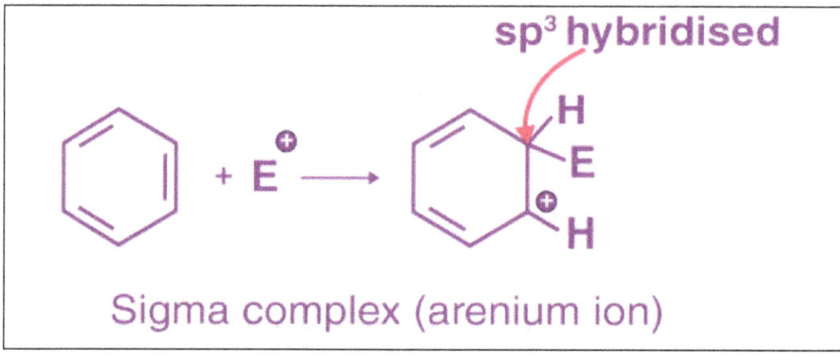

This arenium ion finds stability in a resonance structure. Since the delocalization of electrons stops at the sp3 hybridized carbon, the sigma complex or the arenium ion loses its aromatic character.

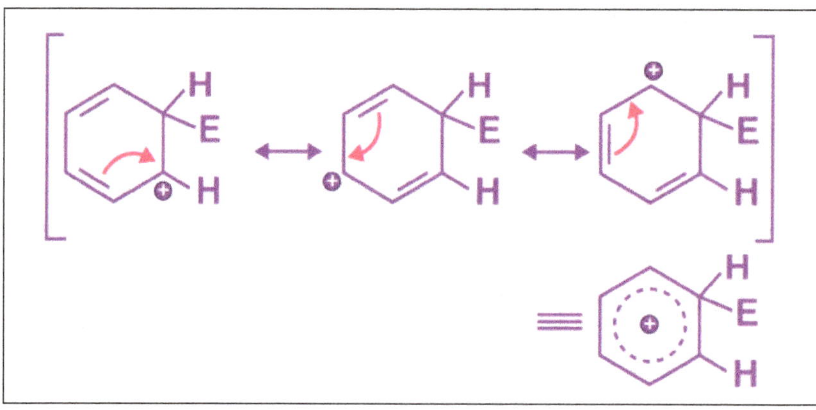

Step 3: Removal of Proton

In order to restore the aromatic character, the sigma complex releases a proton from the sp^3 hybridized carbon when it is attacked by the $[AlCl_4]^-$. The reaction describing the removal of a proton from the sigma complex is given below:

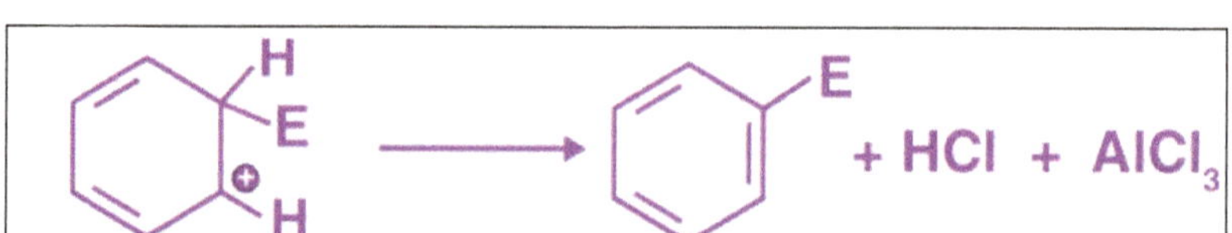

Thus, the electrophile replaces the hydrogen atom in the benzene ring. The electrophilic substitution reaction is a very important reaction in organic chemistry as the concept is used in many organic name reactions.

E1 and E2

The E1 and E2 eliminations present a fair degree of regioselectivity and stereo-selectivity which are governed by their mechanistic aspects.

E1 elimination	E2 elimination
Rate = k [substrate] Unimolecular	Rate = k [substrate][Base] Bimolecular
Leaving group first departs and carbocation is formed from which removal of H takes place.	Concerted process; both the groups (proton and leaving group) are eliminated simultaneously.
Preferred in 3° >2° >1° substrates.	No such preference.
No strong base required.	Stronger and bulky base in high concentration is required.
Strongly dependent on leaving group. Hydroxyl can be converted to a better leaving group via protonation by use of an acid.	Hydroxyl can never be a leaving group for E2 elimination.
No stereochemical requirement.	The leaving group must be trans to the Hydrogen involved.

Regio-Chemical Aspects of E1 and E2 Eliminations: Orientation of the Double Bond

During elimination reaction, a double bond is generated. So, one is concerned with the regiochemistry i.e., the production of structural isomers which have the double bond in different positions. Elimination reactions may be either regiospecific or regioselective. Regiospecific elimination reactions produce only one isomer of an alkene. Regioselective elimination reactions, on the other hand, produce several different isomers, but give one isomer in major quantity than the others. This, in turn, depends on whether the reaction prefers to eliminate only one particular β-hydrogen or a mixture of β-hydrogens.

If an elimination reaction produces the more highly substituted alkene, the reaction is said to follow the Saytzeff rule. On the other hand, if the elimination reaction forms the less substituted alkene, the reaction is said to follow the Hofmann rule. A number of factors including the nature of substrate, the leaving group, reaction conditions etc. determine what rule a given reaction may follow.

Regiochemistry of E1 Elimination

E1 eliminations are regioselective and generate the alkene that has the more substituents, as the major product. Thus, the E1 reactions follow Saytzeff rule. This is due to the fact that in E1 eliminations, leaving group departs to produce first a carbocation. Thus the direction of the double bond is determined almost entirely by the relative stabilities of the olefins produced via elimination. The major product is the alkene that has the more substituents, because this alkene is the more stable of the two possible products. For example, 2-bromo-3-methylbutane undergoes E1 elimination to yield two alkene products viz. 3-methyl-1-butene and 2-methyl-2-butene. The major product is 2-methyl-2-butene because there are three alkyl substituents attached to the carbon—carbon double bond.

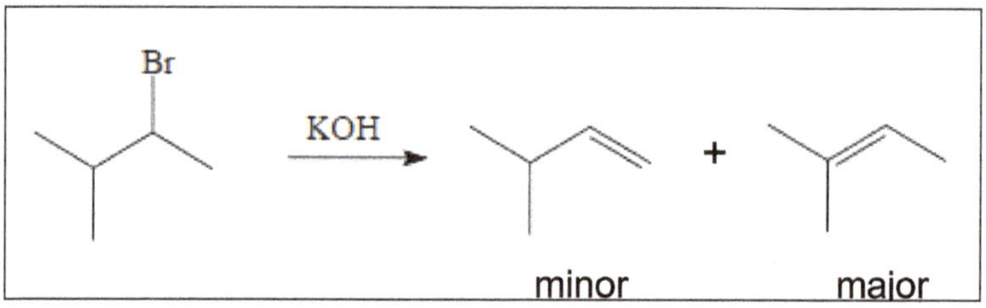

More substituted alkene is more stable' does not necessarily explain why it is the one that forms faster. For this, we need to have a look on the transition state leading to the two alkenes. In the carbocation there is hyperconjugation involving each β-hydrogen. Since the hyperconjugation structures possess some double-bond character, the interaction with hydrogen is greatest at more highly substituted carbons; that is, there will be greater weakening of C–H bonds and more double-bond character at more highly substituted carbon atoms. This structural effect in the carbocation intermediate governs the direction of elimination and leads to the preferential formation of the more highly substituted alkene. More stable transition state is lower in energy, and results in faster formation of more substituted alkene, i.e. it is the thermodynamic product.

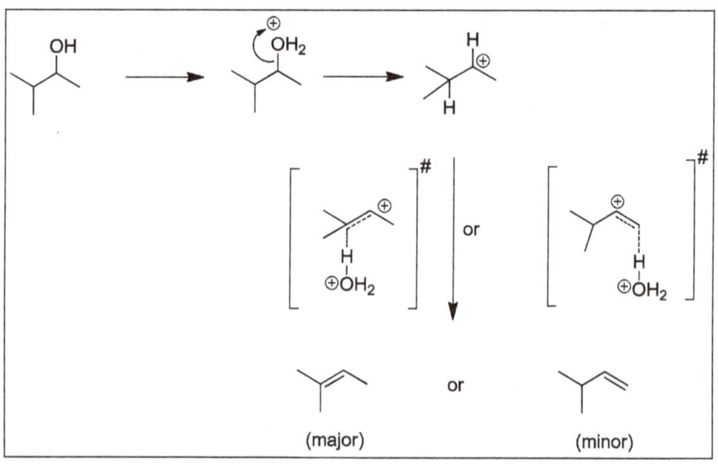

Regiochemistry of E2 Elimination

Most of the E2 reactions, involving good leaving groups (halides, sulfonates etc.) also follow the Saytzeff rule, i.e. lead to a more substituted alkene. This can be attributed to a similar transition state as discussed in the case of E1 elimination. Hence, halides (Br, Cl, I) give more substituted alkene, but E2 elimination of alkyl fluorides results in the least substituted alkene. When a base begins to abstract a proton from an alkyl fluoride, the fluoride ion has less tendency to leave compared to other halide ions as it is a poor leaving group. Therefore negative charge develops on the carbon that is losing the proton, giving the transition state a carbanion character (rather than an alkene character as shown previously) which is stabilized by highly electron-withdrawing fluorine and fewer electron-donating alkyl substituents. This subsequently results in the formation of least-substituted alkene.

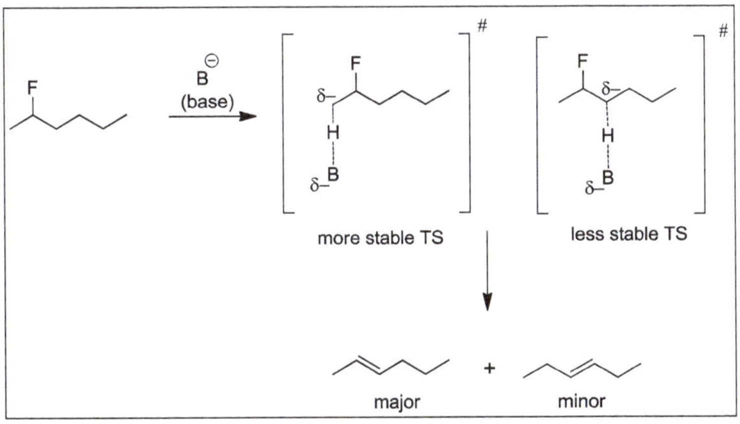

Similarly, E2 reactions involving other poor leaving groups, particularly those with quaternary ammonium salts, are said to follow the Hofmann rule and give primarily the less-substituted alkene. Moreover, when the proton to be removed is in the sterically more hindered position or when the base used for E2 elimination is a bulky base, exceptions to the Saytzeff rule occur. These situations lead to a high proportion of the less substituted alkene (Hofmann orientation).

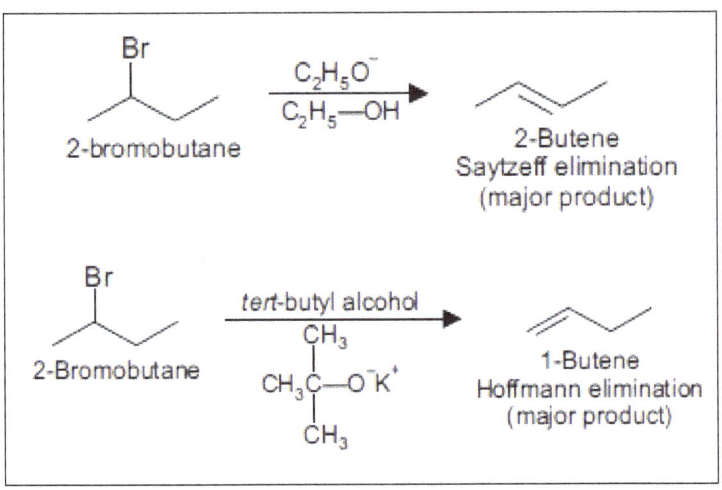

However, when formation of a double bond can lead to extended conjugation with already present functional groups (C=O or C=C) or aryl ring, then irrespective of the mechanism, the conjugated product only predominates.

Similarly, eliminations in bridged cyclic compounds follow the Bredt rule, i.e. irrespective of the mechanism, a double bond never goes to the bridgehead carbon in rings of smaller size. This is simply a consequence of the strain induced by a planar bridgehead carbon, and so having a double bond on a bridgehead would be too unstable.

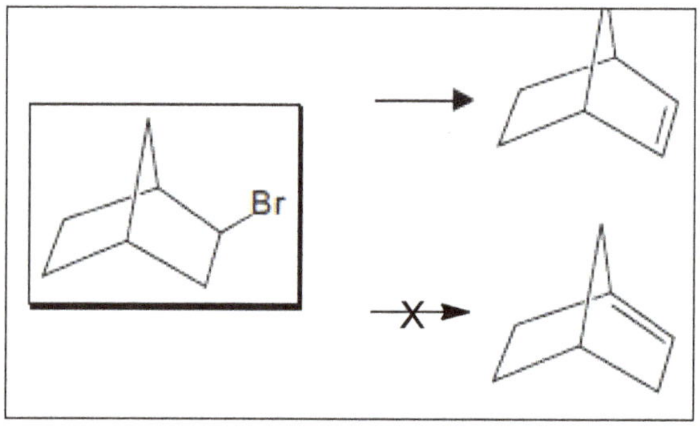

Stereochemical Aspects of E1 and E2 Elimination Mechanisms

The E1 reaction mechanism is a two-step process that, as with the S_N1 mechanism, usually loses all the stereochemical information of the substrate as the reaction proceeds. On the other hand, the E2 mechanism is a concerted mechanism. A concerted reaction usually requires that the substrate have a specific conformation that allows the orbitals of the bonds being broken to overlap the bonds being formed. It is only when the orbitals overlap, the electrons flow smoothly from the breaking bonds to the forming bonds and consequently the reaction takes place. These aspects of elimination reactions are discussed as follows:

Stereochemistry of E1 Eliminations

Considering the geometrical isomerism of the E1 elimination product, usually the formation of Ealkene is favoured over the Z-alkene. This is due to the steric factors of the transition states leading to these possible geometries, favouring E-geometry because the substituents can get farther apart from each other. So, E-alkene forms faster.

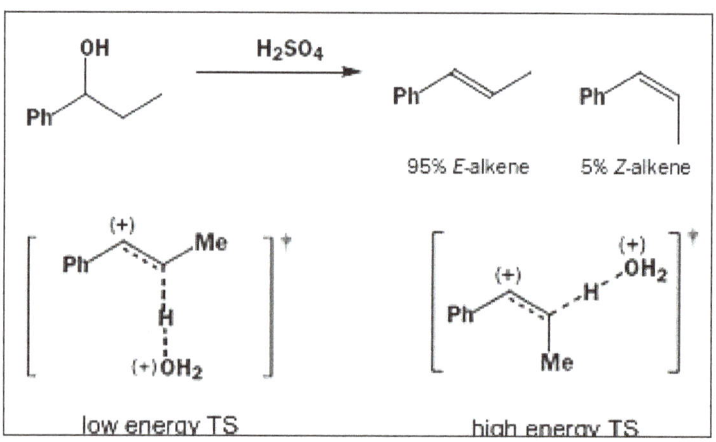

References

- General-introduction-organic-compounds, organic-chemistry: toppr.com, Retrieved 10, May 2020
- Classification-of-organic-compounds, purification-classification-and-nomenclature-of-organic-compounds: askiitians.com, Retrieved 18, July 2020
- Classification-organic-compounds, organic-chemistry: toppr.com, Retrieved 24, January 2020
- Types-of-organic-reactions: byjus.com, Retrieved 06, April 2020
- Electrophilic-substitution-reaction-mechanism: byjus.com, Retrieved 15, June 2020

4

Inorganic Chemistry

The branch of chemistry which focuses on inorganic and organometallic compounds is termed as inorganic chemistry. It studies their synthesis and behavior. The topics elaborated in this chapter will help in gaining a better perspective about inorganic chemistry, inorganic compounds, inorganic reactions and redox reactions.

The branch of chemistry that deals with the study of compounds, which does not consist of carbon-hydrogen atoms in it, is called 'Inorganic Chemistry.' In simple words, it is opposite to that of Organic Chemistry. The substances which do not have carbon-hydrogen bonding are the metals, salts, chemical substances, etc.

On this planet, there are known to exist about 100,000 numbers of Inorganic compounds. Inorganic chemistry studies the behaviour of these compounds along with their properties, their physical and chemical characteristics too. The elements of the periodic table except for carbon and hydrogen, come in the lists of Inorganic compounds.

Many of the elements are technologically important: titanium, iron, nickel and copper, for example, are used structurally and electrically. Second, the transition metals form several useful alloys, with each other and with other metallic elements.

Chemistry

Chemistry is the study of the substances of which matter is composed. Chemistry is heavily experimental because you can only study reductions if you totally mix substances together. Luckily, you don't need to risk your health and safety to understand chemistry better because brave scientists already did. There are over a hundred elements that make up the matter of our world and universe. They combine to make thousands and thousands of compounds.

A chemical compound is made up of atoms of different elements joined together by a chemical bond. The bonds are so strong that the compound acts as if it were a single substance. The joined atoms form molecules and the molecules connect together to make the compound.

A chemical formula is a notation used by scientists to show the number and type of atoms present in a molecule using atomic symbols and numeric subscriptions. A chemical formula is a simple representation, in writing, of a three-dimensional molecule that exists. A chemical formula describes a substance, down to the exact atoms that make it up. There are three basic types of chemical formula, the empirical formula, the molecular formula and the structural formula.

Organometallic Chemistry

Organometallic Chemistry, an interdisciplinary science in Inorganic Chemistry, has grown at a phenomenal pace during the last three to four decades. On the academic plane, efforts to elucidate the nature of bonds in the ever increasing list of exciting organometallic compounds have led to a clearer understanding of the nature and variety of chemical bonds. Organometallic compounds are primarily used as homogeneous catalysis agents in industries.

Organometallic chemistry is an organometallic compound study. Because many compounds without these bonds are chemically identical, an alternative may be compounds containing metallic bonds of a mostly covalent nature. Organometallic chemistry blends elements of inorganic chemistry with organic chemistry.

Transition Elements

A transition element may be defined as one which possesses partially filled d-orbitals in its penultimate shell. This conceptual definition is useful as it enables us to recognize a transition element merely by looking at its electronic configuration. This definition excludes zinc, cadmium and mercury from the transition elements as they do not have a partially filled d-orbital. However, they are also considered as transition elements, because their properties are an extension of the properties of transition elements in inorganic chemistry. In fact, the zinc group serves as a bridge between the transition elements and the representative elements.

The most notable characteristics shared by the 24 elements concerned are that they are all metals and that most of them are hard, solid and lustrous, have high melting and boiling points and are good conductors of heat and electricity. The range in these properties is considerable; hence, the statements are comparable to the general properties of all the other elements.

P-Block Elements

The elements placed in group 13 to group 18 of the periodic table constitute the p-block. The properties of inorganic chemistry p block elements like that of other block elements are greatly influenced by their atomic size, ionization enthalpy, electron gain enthalpy and electronegativity. The absence of d–orbitals in the second period and presence of d- or f-orbitals in heavier elements has a significant effect on the properties of the elements and therefore, heavier p-block elements differ from their lighter congeners.

Applications of Inorganic Chemistry

Inorganic chemistry finds its high number of applications in various fields such as Biology, chemical, engineering, etc.

- It is applied in the field of medicine and also in healthcare facilities.
- The most common application is the use of common salt or the compound Sodium hydroxide in our daily lives.
- Baking soda is used in the preparation of cakes and other foodstuffs.
- Many inorganic compounds are utilized in ceramic industries.
- In the electrical field, it is applied to the electric circuits as silicon in the computers, etc.

Inorganic Compounds

Inorganic compound is any substance in which two or more chemical elements (usually other than carbon) are combined, nearly always in definite proportions. Compounds of carbon are classified as organic when carbon is bound to hydrogen. Carbon compounds such as carbides (e.g., silicon carbide [SiC_2]), some carbonates (e.g., calcium carbonate [$CaCO_3$]), some cyanides (e.g., sodium cyanide [NaCN]), graphite, carbon dioxide, and carbon monoxide are classified as inorganic.

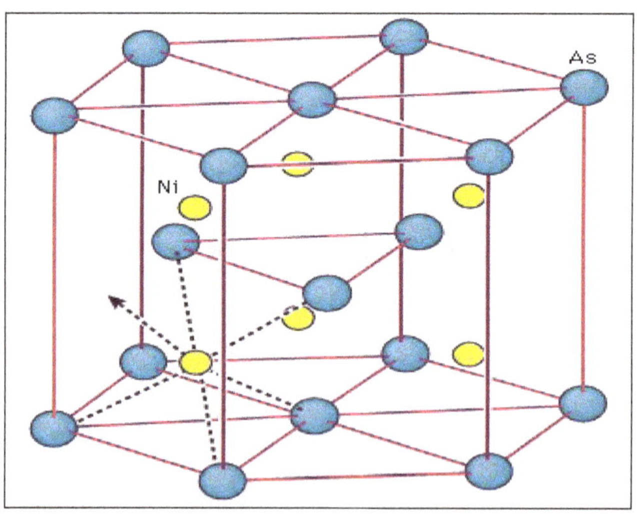

The crystal structure of the inorganic compound nickel arsenide. This type of structure departs strongly from that expected for ionic bonding and shows the importance of covalence. There is also some direct nickel-nickel bonding that tends to draw the nickel atoms together.

Classification of Inorganic Compounds

The organic compounds that are classified under Inorganic chemistry are:

- Acids Acids are those compounds that dissolve in water and generate hydrogen ions or H^+ Ions. The examples of acids include Hydrochloric acid, citric acid, sulphuric acid, vinegar, etc. One example of the acidic reaction is shown below:

  Hydrochloric acid + water → H^+ + Cl

- Bases A base is a type of substance or a compound that produces hydroxyl ions when kept in water. The bases like potassium hydroxide, calcium hydroxide, ammonia, sodium hydroxide produce OH^- ions when dissolved in water.

  Potassium Hydroxide + H_2O → K^+ + OH^-

- Salts As you might be familiar with the word 'Salt'. The substances obtained as a result of the reaction between an acid and a base are called Salts. The table salt of sodium hydroxide is one of the typical examples of salts.

- Oxides he compounds which consist of one oxygen atom called Oxides.

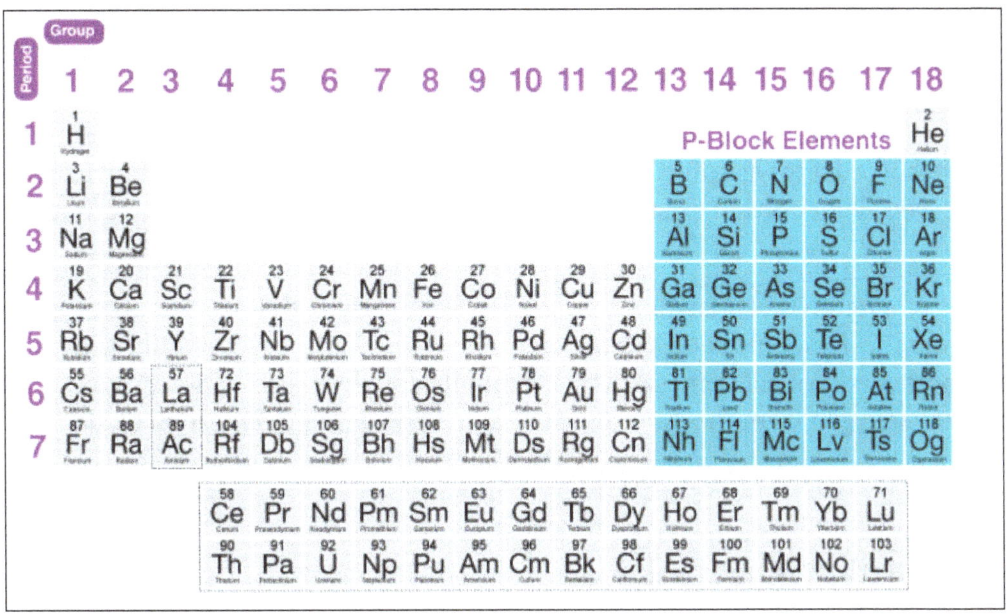

Inorganic Reactions

Many chemical reactions can be classified as one of five basic types. Having a thorough understanding of these types of reactions will be useful for predicting the products of an unknown reaction. The five basic types of chemical reactions are combination, decomposition, single-replacement, double-replacement, and combustion. Analyzing the reactants and products of a given reaction will allow you to place it into one of these categories. Some reactions will fit into more than one category. Identifying a reaction as one of these types does not preclude it from also being an oxidation-reduction (redox) reaction. For example, many combination reactions can also be classified as a redox reaction.

Combination Reactions

A combination reaction, also known as a synthesis reaction, *is a reaction in which two or more substances combine to form a single new substance.* Combination reactions can also be called synthesis reactions .The general form of a combination reaction is:

$$A + B \rightarrow AB$$

One example of a combination reaction is two elements combining to form a compound. Solid sodium metal reacts with chlorine gas to product solid sodium chloride.

$$2Na(s) + Cl_2(g) \rightarrow 2NaCl(s)$$

This reaction can also be classified as a redox reaction due to the changes in oxidation states. Sodium goes from a 0 to +1 oxidation state while chlorine goes from a 0 to a −1 oxidation state.

Notice that in order to write and balance the equation correctly, it is important to remember the seven elements that exist in nature as diatomic molecules ($H_2, N_2, O_2, F_2, CI_2, Br_2$ and I_2).

One sort of combination reaction that occurs frequently is the reaction of an element with oxygen to form an oxide. Metals and nonmetals both react readily with oxygen under most conditions. Magnesium reacts rapidly and dramatically when ignited, combining with oxygen from the air to produce a fine powder of magnesium oxide.

$$2Mg(s) + O_2(g) \rightarrow 2MgO(s)$$

Decomposition Reactions

A decomposition reaction is a reaction in which a compound breaks down into two or more simpler substances. The general form of a decomposition reaction is:

$$AB \rightarrow A + B$$

Most decomposition reactions require an input of energy in the form of heat, light, or electricity. Binary compounds are compounds composed of just two elements. The simplest kind of decomposition reaction is when a binary compound decomposes into its elements. Mercury (II) oxide, a red solid, decomposes when heated to produce mercury and oxygen gas.

$$2HgO(s) \rightarrow 2Hg(l) + O_2(g)$$

A reaction is also considered to be a decomposition reaction even when one or more of the products is still a compound. A metal carbonate decomposes into a metal oxide and carbon dioxide gas. For example, calcium carbonate decomposes into calcium oxide and carbon dioxide.

$$CaCO_3(s) \rightarrow CaO(s) + CO_2(g)$$

Metal hydroxides decompose on heating to yield metal oxides and water. Sodium hydroxide decomposes to produce sodium oxide and water.

$$2NaOH(s) \rightarrow Na_2O(s) + H_2O(g)$$

Single-Replacement Reactions

A single-replacement reaction, also known as single-displacement or substitution reaction, is a reaction in which one element replaces a similar element in a compound. The general form of a single-replacement reaction is:

$$A + BC \rightarrow AC + B$$

In this general reaction, element A is a metal and replaces element B, also a metal, in the compound. When the element that is doing the replacing is a nonmetal, it must replace another nonmetal in a compound, and the general equation becomes:

$$Y + XZ \rightarrow XY + Z$$

Y is a nonmetal and replaces the nonmetal Z in the compound with X.

Magnesium is a more reactive metal than copper. When a strip of magnesium metal is placed in an aqueous solution of copper (II) nitrate, it replaces the copper. The products of the reaction are aqueous magnesium nitrate and solid copper metal.

$$Mg(s) + Cu(NO_3)_2(aq) \rightarrow Mg(NO_3)_2(aq) + Cu(s)$$

Many metals react easily with acids, and, when they do so, one of the products of the reaction is hydrogen gas. Zinc reacts with hydrochloric acid to produce aqueous zinc chloride and hydrogen (see figure below).

$$Zn(s) + 2HCl(aq) \rightarrow ZnCl_2(aq) + H_2(g)$$

Zinc metal reacts with hydrochloric acid to give off hydrogen gas in a single-replacement reaction.

Pictured here is about 3 pounds of sodium metal reacting with water. Sodium metal reacts vigorously when dropped into a container of water, giving off hydrogen gas. A large piece of sodium will often generate so much heat that the hydrogen will ignite.

Double-Replacement Reactions

A double-replacement reaction, also known as double-displacement, is a reaction in which the positive and negative ions of two ionic compounds exchange places to form two new compounds. The general form of a double-replacement reaction is:

$$AB + CD \rightarrow AD + CB$$

In this reaction, A and C are positively-charged cations, while B and D are negatively-charged anions. Double-replacement reactions generally occur between substances in aqueous solution. In order for a reaction to occur, one of the products is usually a solid precipitate, a gas, or a molecular compound such as water.

A precipitate forms in a double-replacement reaction when the cations from one of the reactants combine with the anions from the other reactant to form an insoluble ionic compound. When aqueous solutions of potassium iodide and lead (II) nitrate are mixed, the following reaction occurs.

$$2KI(aq)+Pb(NO_3)_2(aq)\rightarrow 2KNO_3(aq)+PbI_2(s)$$

When a few drops of lead (II) nitrate are added to a solution of potassium iodide, a yellow precipitate of lead (II) iodide immediately forms in a double-replacement reaction.

Combustion Reactions

A combustion reaction is a reaction in which a substance reacts with oxygen gas, releasing energy in the form of light and heat. Combustion reactions must involve O_2 as one reactant. The combustion of hydrogen gas produces water vapor.

$$2H_2(g)+O_2(g)+2H_2O(g)$$

Notice that this reaction also qualifies as a combination reaction.

Figure shows the Hindenburg was a hydrogen-filled airship that suffered an accident upon its attempted landing in New Jersey in 1937. The hydrogen immediately combusted in a huge fireball, destroying the airship and killing 36 people. The chemical reaction was a simple one: hydrogen combining with oxygen to produce water.

Many combustion reactions occur with a hydrocarbon, a compound made up solely of carbon and hydrogen. The products of the combustion of hydrocarbons are always carbon dioxide and water. Many hydrocarbons are used as fuel because their combustion releases very large amount of heat energy. Propane (C_3H_8) is a gaseous hydrocarbon that is commonly used as the fuel source in gas grills.

$$C_3H_8(g) + 5O_2(g) \rightarrow 3CO_2(g) + 4H_2O(g)$$

Redox Reactions

In redox reactions, one element or compound is reduced (gains electrons) and another is oxidized (loses electrons). In terms of everyday life, redox reactions occur all of the time around us. For example, the metabolism of sugars to CO_2, which stores energy in the form of ATP, is a redox reaction. Another example of redox is fire or combustion, such as in a car engine. In a car engine, hydrocarbons in the fuel are oxidized to carbon dioxide and water, while oxygen is reduced to water. Corrosion (i.e. the formation of rust on iron) is a redox reaction involving oxidation of a metal.

Oxidation states of vanadium in acidic solution. From left to right the oxidation state goes from +5 to +2. These four oxidation states form the basis of the vanadium flow battery, a storage device for electricity generated from sunlight and wind.

Oxidation-reduction reactions are important to understanding inorganic chemistry for several reasons:

- Transition metals can have multiple oxidation states.
- Main group elements (N, halogens, O, S...) also have multiple oxidation states and important redox chemistry.
- Many inorganic compounds catalyze redox reactions (which are especially useful in industrial and biological applications).

- Energy conversion and storage technologies (solar water splitting, batteries, electrolyzers, fuel cells) rely on inorganic redox reactions and catalysis.
- Electrochemistry provides a way to measure equilibrium constants for dissolution/precipitation, complexation, and other reactions.
- Reaction mechanisms in organometallic chemistry (oxidative addition, reductive elimination) involve changes in the oxidation states of metals.

Redox reaction could occur by the direct transfer of electrons (as in some electrochemical cells and in many solution reactions) or by the transfer of atoms and ions. In cases of coordination complexes, two reacting complexes could come together by a bridging ligand so electron transfer could also occur through this one. But it may not be the case in all types of redox reactions i.e. it is not mandatory that for redox reaction to occur, there must be an involvement of bridging ligand. It means that we will be discussing different types of mechanisms to discuss redox reactions.

Because redox reactions in solution involve both an oxidizing and a reducing agent, they are usually bimolecular in character. But in few exceptional cases one molecule has both oxidizing and reducing centres.

In the 1950s, Prof. Henry Taube who received Nobel Prize in Chemistry (1983) "his work in the mechanisms of electron-transfer reactions, especially in metal complexes." He identified two different mechanisms of redox reactions for metal complexes:

(a) Inner-Sphere Mechanism: In an inner-sphere mechanism, the coordination spheres of the reactants share a ligand transitorily and form a bridged transition state.

(b) Outer-Sphere Mechanism: In an outer-sphere mechanism, the complexes come into contact without sharing a bridging ligand and the electron tunnels from one metal atom to the other.

Although the mechanisms of some redox reactions have been definitively assigned as inner- or outersphere.

The mechanisms of a vast number of reactions are unknown because in order to study or figure out the reaction pathway there should be sufficient time. But in case of labile complexes (According to Taube, a reaction half-life (the time it takes for the concentration of the initial compound to decrease by one half) of one minute or less as the criterion for lability). So where the half life is less than one minute it would be difficult to make unambiguous assignments that whether the electron transfer takes place through inner-sphere or outer-sphere mechanism.

Redox Reactions: Electron Transfer Reactions

The redox reactions are basically electron transfer reactions can be chemically defined as 'It is the study and manipulation of valence electrons of atoms or molecules'. In principle, redox reactions are considered as electron-transfer reactions however mechanistic studies have demonstrated that in some of the redox reactions transfer of atoms or ions also takes place. Redox Reactions needs straightforward transfer of electrons such as a) electron transfer takes place from one metal to another in electrochemical cell, b) transfer of group of ligands along with their electrons that can efficiently oxidize or reduce a metal centre. Prior to define the concept, one must understand the meaning of oxidation and reduction. Redox reaction literally visualizes the transfer of following species between two redox partners:

- Electron(s)
- Hydrogen Atom
- Hydride Ion
- Oxygen Atom
- Chlorine Atom etc.

In general, we consider redox reactions as electron-transfer which takes place as:

$$A + B- \rightarrow A^{-} + B$$

In a redox reaction oxidation and reduction both occurs simultaneously nevertheless we have to define both oxidation and reduction separately. It is possible by numerous ways to define word oxidation although the most prominent definitions include:

- Oxygen addition.
- Hydrogen removal.
- Transfer of electron(s).

To mechanism of a redox reaction can be understood if one clearly finds out whether the transfer of an atom, a group or an electron has occurred during the course of reaction. In addition, it is also necessary to know which atom(s) is/are transferred or how many electrons are transferred along with what kind of the transition states of all steps. If an oxidant accepts one electron, it is considered equivalent to the acceptance of a hydrogen atom and if an oxidant accepts a couple of electrons, it is considered equivalent to the acceptance of a hydride ion. The thermodynamic features of the redox reactions are of basic importance but abovementioned information can be obtained by kinetic measurements.

The important information on the mechanisms of the electron-transfer reactions will be obtained by methodical kinetic studies of the reductions of the cobalt (III) complexes in the organic solvents and in the organic solvent-water mixtures. These studies were carried out because the complexes of cobalt in the oxidized form (Co^{3+}) are inert whereas labile in the reduced form (Co^{2+}). Likewise, the kinetic studies on Cr(III)/Cr(II) (πd^3 / $\pi d^3 \sigma d^1$) couple in an octahedral systems can be investigated because the σ electron in Cr(II) is antibonding in nature which is accountable to the tetragonal distortion in Cr(II) complexes as well as their vast labile character relative to those of Cr(III) complexes. It should be noted that, Ru(III)/Ru(II) (πd^5 /πd^6) redox couples in a low spin octahedral complexes exhibit quite opposite behaviour since Ru(II) systems in this case is intert in comparison to Ru(III). Here, it is noteworthy to mention that in contrast to the electrochemical reactions (involve charge transfer heterogeneously), in a homogeneous reaction both the donor and acceptor are in the same phase (solution) therefore the mechanism basically involves the concerned participating species in the solution.

Redox Reactions in Transition Metal Complexes

The redox reactions or electron-transfer reactions of metal complexes are one of the fundamental study subjects in inorganic and/or bioinorganic chemistry. Reaction between two metal complexes leading to the change in complementary oxidation states and localized on the metal centres have been investigated and studied comprehensively. Henry Taube observed that transfer of an atom from one metal to another usually needs entrance of first metal the coordination sphere of second metal ion through either a bridging ligand or by ligand substitution. Conversely, some electron transfer can also be resulted keeping the primary coordination sphere of any reactant unchanged because the electrons can cross the barriers like ligands.

Under two main headings, a total of four sets of redox reactions are recognized i.e. outer sphere electron transfer reactions (direct electron transfer) and inner sphere electron transfer reactions (atom/group/ion transfer); including (1) Inorganic oxidant and inorganic substrate, (2) Inorganic oxidant and organic substrate, (iii) Organic oxidant and organic substrate and, (iv) Organic oxidant and inorganic substrate. The redox reactions concerning metal ions can be categorized on the basis of species transfer; whether the electron transfer incorporates ionic sphere or it involves coordination sphere of the metal complex. An example of metal complex based redox reaction is given below:

$$\begin{array}{ccccccc} MnO_4^- & + & MnO_4^{2-} & \rightleftharpoons & MnO_4^{2-} & + & MnO_4^- \\ & & & & \downarrow & & \downarrow \\ & & & & BaMnO_4 & & Ph_4AsMnO_4\,(ppt) \end{array}$$

If the redox reaction involves ionic sphere only it is considered as direct electron transfer reactions where the transfer of electron takes place between two species without changing any other parameters. While the species are metal complexes such type of reactions are called outer sphere reactions. The characteristic features of outer sphere reactions are:

- These reactions exclude any bridging of ligands between two complexes acting as redox couple.

- The metal complex redox couples simply function as electron donor (oxidation) and electron acceptor (reduction) systems. Such as:

$$[M_1L_n]^{x-} + [M_2L_n]^{y-} \rightarrow [M_1L_n]^{(x-1)-} + [M_2L_n]^{(y+1)-}$$

Where, M = metal; n = 1,2,3,4...etc; x or y = charge.

Scheme 1

The outer sphere mechanism has paid special attention and being quite important due to the profound similarity of electron transfer in metal complexes with the electron transfer process in metalloenzymes. Theoretically, transfer of electron occurs from reductant to oxidant without causing any modification in the coordination spheres of each redox couple in the outer sphere mechanism. Such redox reactions are very fast despite electron transfer takes place between two substitution-inert complexes.

For example:

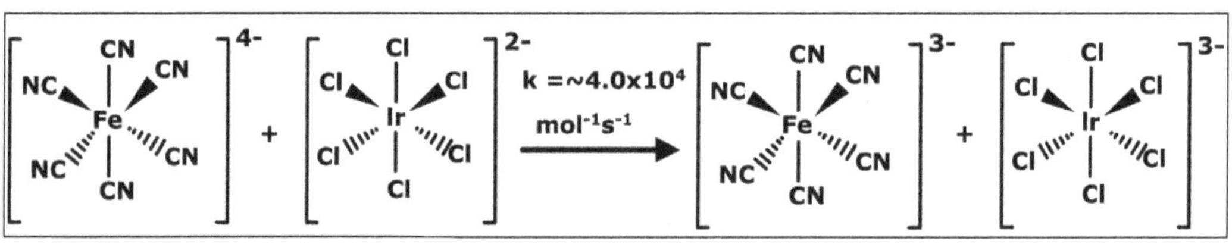

It is evident from the above reaction that except electron transfer from $[Fe(CN)6]^{4-}$ to $[IrCl6]^{2-}$, none of the other parameters changes.

On the other hand, if the redox reaction includes coordination sphere, it is considered as atom, group or ion transfer reactions where the transfer of atom/group/ion takes place between two redox couples. While the species are metal complexes such type of reactions are called inner sphere reactions. The characteristic features of inner sphere reactions are:

- These reactions require the formation of bridged bimetallic complexes.
- In these reactions, ligand transfer takes place at the same time.

$$[M_1L_n]^x + [XM_2L_{m-1}]^y \longrightarrow [L_{n-1}M_1 - X - M_2L_{m-1}]^{x+y} + L$$
$$\downarrow nL$$
$$[L_{n-1}M_1 - X]^y + [M_2L_n]^x$$

Where, n or m = metal; n = 1, 2, 3, 4...etc;

X= Anionic ligand i.e. CI^-, NCS^- etc.

Scheme 2

The inner sphere mechanism involves sharing of the ligand between reductant and oxidant in their inner coordination spheres and the transfer of the electron takes place across the bridging group. According to the electrochemical and structural studies, the redox stability/reactivity of transition metal complexes may be determined by the position of ligands in the coordination sphere pertaining to the redox orbital symmetry, along with the electronic nature of ligands.

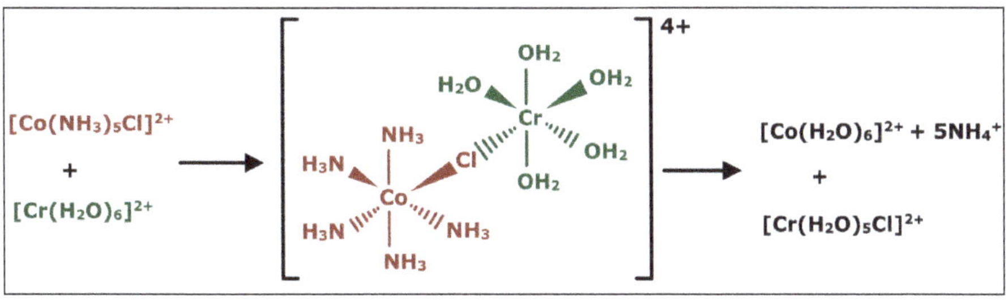

The transfer of electron will result in the change in the total optical rotation therefore rate of racemisation will be equal to the rate of electron transfer provided reagents used are taken in equal concentrations.

While the metal ion being Cr(II) it acts as reducing agent and favours an inner sphere reaction. The bridging Cl- moves towards Cr(II) from Cr(III) by lowering the energy of its acceptor orbital and simultaneously raising the energy of Cr(II) donor orbital. However, such a nuclear motions and such a correlation phenomenon for electron transfer are lacking for an outer sphere mechanism.

Types of Redox Reactions

One: One Equivalent Reaction

These are simple one electron transfer reactions which act as models to understand the processes involved in similar reaction containing more complicated species. The key features i.e. electron donor and electron acceptor sites of those complicated systems are correlated with simple one to one equivalent redox reactions to follow mechanistic pathways.

The examples for such reactions are:

- $Ce^{3+} + Co^{3+} \rightarrow Co^{2+} + Ce^{4+}$
- $Fe^{2+} + Co^{3+} \rightarrow Fe^{3+} + Co^{2+}$

The kinetic data for oxidation and reduction of metal ions can be obtained from the appearance/disappearance of the M^{n+} in the absorption spectrum. In the above first reaction, Co^{3+} disappears from the solution which is evident from the absence of absorption maxima at 605m. it strongly support the reduction of Co^{3+} to Co^{2+}. Likewise, kinetic data on oxidation of Fe^{2+} in perchloric acid by Co^{3+} can be acquired by making the concentrations of both the reactants equal.

Two: One Equivalent Reaction

It is the kind of reactions involves reactants in 2:1 stoichiometry and several such reactions are recognized. For instance, reaction of Fe^{2+} (one equivalent) with oxidants (two equivalent), form complex wherein, either Fe^{2+} or oxidants undergo in unstable state. The reaction follows as:

$$TI^{3+} + Fe^{2+} \underset{K_2}{\overset{K_1}{\rightleftharpoons}} TI^{2+} + Fe^{3+}$$

$$TI^{2+} + Fe^{3+} \underset{K_4}{\overset{K_3}{\rightleftharpoons}} TI^{+} + Fe^{3+}$$

The above reaction couple incorporates two consecutive one electrons transfer where oxidation takes place at Fe^{2+} whereas reduction at Tl^{2+} in $HClO_4$ (aq).

Another redox couple (V^{4+}/ Tl^{3+}) for oxidation of V4+ by Tl3+ exhibit analogous reaction mechanism to that for Fe^{2+} and Tl^{2+} shown above. Moreover, following reaction mechanism involves oxidation of Tl^{+} by Ce^{4+} in 6.2M HNO3 at 54 °C.

$$2Ce^{4+} \rightleftharpoons \left[Ce^{4+}\right]_2$$

$$Ce^{3+} + Ce^{4+} \rightleftharpoons \left[Ce^{3+} + Ce^{4+}\right]$$

$$Ce^{4+} + OH^{-} \rightleftharpoons Ce^{3+} + OH$$

$$TI^{+} + OH \rightleftharpoons TI^{2+} + OH^{-}$$

$$TI^{+} + Ce^{4+} \rightleftharpoons TI^{2+} + Ce^{3+}$$

$$TI^{2+} + Ce^{4+} \rightleftharpoons TI^{3+} + Ce^{3+}$$

Two: Two Equivalent Reactions

These reactions involve both the reactants in two equivalents however the equivalency is considered not in moles but the transfer of electron(s) between reacting species. A fascinating example is depicted below:

$$U^{4+} + TI^{3+} \longrightarrow U^{6+} + TI^{+}$$

The mechanism for the above reaction is:

$$U^{4+} + TI^{3+} \longrightarrow U^{5+} + TI^{2+}$$

$$U^{4+} + TI^{2+} \longrightarrow U^{5+} + TI^{+}$$

$$U^{5+} + TI^{3+} \longrightarrow U^{6+} + TI^{2+}$$

Complementary Redox Reactions

If the oxidants and reductants both alter their oxidation state by equal in number, such reactions are called complementary redox reactions. These reactions are important because only one molecule of the each reactant is needs to react. For instance:

$$Sn^{2+} + TI_{3+} \longrightarrow Sn^{4+} + TI^{+}$$

$$Sn^{2+} + Hg^{2+} \longrightarrow Sn^{4+} + Hg^{o}$$

Non-Complementary Redox Reactions

Unlike to the complementary reactions, when the oxidants and reductants alter their oxidation number by unequally, such reactions are known as non-complementary reactions. Therefore, the number of molecules of the reacting species may be different with respect to each other. An example of such reaction is given below:

$$2Fe^{3+} + Sn^{2+} \longrightarrow 2Fe^{2+} + Sn^{4+}$$

Mechanism of Redox Reactions: An Outer Sphere Reaction

The metal complexes which do not go through substitution reaction involve outer sphere electron transfer reaction. It means none of the bonds are broken or formed in either of the metal complex reactants. In other words, these reactions include complementary change in the oxidation state of both the metals through a pathway where the ligand does not undergo "bridging" mode in the coordination sphere of the complexes. Therefore, coordination spheres of the metal complex reactants in an outer-sphere reaction stay intact. The favourable factor for an outer sphere reaction is the inertness of the complexes towards substitution over the time needed for a redox reaction to occur.

$$\underset{\text{Oxident}}{M^{(a+1)+}L_x} + \underset{\text{reductant}}{M^{b+}L_y} \rightleftharpoons \underset{\text{reduced}}{M^{a}L_x} + \underset{\text{oxidized}}{M^{(b+1)+}L_y}$$

Examples of outer sphere reaction:

$$\left[Fe^{II}(CN)6\right]^{4-} + \left[Ir^{IV}CI_6\right]^{2-} \longrightarrow \left[Fe^{III}(CN)_6\right]^{3-} + \left[Ir^{III}CI_6\right]^{3-}$$

$$\left[Fe(Me_2bipy)_3\right]^{2+} + \left[Fe(bipy)_3\right]^{3+} \rightleftharpoons \left[Fe(Me_2bipy)_3\right]^{3+} + \left[Fe(bipy)_3\right]^{2+}$$

$$Rate = k\left[Fe(Me_2bipy)_3\right]^{2+} + \left[Fe(bipy)_3\right]^{3+}$$

These reactions are genuinely quite fast and before L_x or L_y exchange or bridging occurs, redox process takes place.

The coordination sphere of the complex remains intact because reactants do not interact significantly with each other during the electron transfer; i.e. through space, diffusion control.

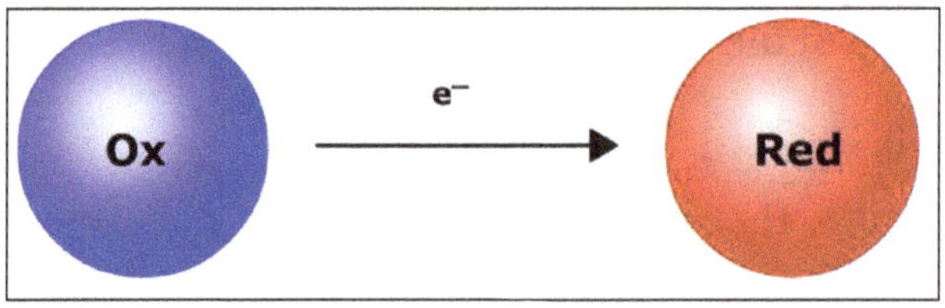

Examples of such behaviour are "self-exchange" electron transfer reactions between homonuclear complexes.

$$[ML_6]^{2+} + [ML_6]^{3+} \longrightarrow [ML_6]^{3+} + [ML_6]^{2+}$$

The outer sphere rates for self-exchange reactions are:

M (metal)	L_n (Ligand)	d^n, spin states	k ($L\ mol^{-1}\ s^{-1}$)
Fe	$(Bipy)_3$	d^5 LS, d^6 LS	$>10^6$
Fe	$(H_2O)_6$	d^5 HS, d^6 HS	3
Os	$(Bipy)_3$	d^5 LS, d^6 LS	$>10^6$
Co	$(Phen)_3$	d^6 LS, d^7 HS	40
Co	$(en)_3$	d^6 LS, d^7 HS	$>10^{-4}$
Co	$(NH_3)_6$	d^6 LS, d^7 HS	$>10^{-6}$

The representative example of self-exchange reaction is:

$$\left[Ru\left(NH_3\right)_6\right]^{2+} + \left[Ru^*\left(NH_3\right)_6\right]^{3+} \xrightarrow{k_{11}=8.2\times10^2\,M^{-1}s^{-1}} \left[Ru\left(NH_3\right)_6\right]^{3+} + \left[Ru^*\left(NH_3\right)_6\right]^{2+}$$

$$\left[Ru^*\left(NH_3\right)_6\right]^{3+/2+} \quad E^o = 0.051v$$

Wherein, Ru and Ru* are isotopes ($\Delta G^o = 0$)

Other similar self-exchange reactions are:

$$\left[Co^*\left(H_2O\right)_6\right]^{2+} + \left[Cr\left(H_2O\right)_6\right]^{3+} \xrightarrow{k_{22}\leq 2.5\times10^{-5}\,M^{-1}s^{-1}} \left[Co^*\left(H_2O\right)_6\right]^{3+} + \left[Cr\left(H_2O\right)_6\right]^{2+}$$

$$\left[Co^*\left(NH_3\right)_6\right]^{3+/2+} \quad E^o = 0.058v$$

Despite odd situation and requirements for an outer sphere reaction, the lowest energy barrier notwithstanding can be yet presented by an outer sphere pathway. This criterion ambiguity is observed in metal-aquo complexes in which one of these is labile in terms of substitution over redox reaction timescale. For instance,

$$\left[Cr^{*}\left(H_2O\right)_6\right]^{2+} + \left[Cr\left(H_2O\right)_6\right]^{3+} \xrightarrow{k_{22} \leq 2.5\times10^{-5} M^{-1}s^{-1}} \left[Cr^{*}\left(H_2O\right)_6\right]^{3+} + \left[Cr\left(H_2O\right)_6\right]^{2+}$$

For such reactions, indirect arguments may be raised for a while i.e. outer sphere rate of reaction for oxidation of $[Cr(H_2O)6]^{2+}$ complex appears to be typically 10-60 times slower relative to the well-known outer sphere reaction involving $[Cr(H_2O)6]^{2+}$ employing same oxidant.

The relative rates of self-exchange reactions for $[Cr(H_2O)_6]^{3+}/[Cr(H_2O)_6]^{2+}$ redox couple vs $[(NH_3)5RuH_2O]^{3+}/(NH_3)5RuH_2O]^{2+}$ couple is fairly informative because the higher oxidation state (3+ charged complex) for each couple is transformed to the hydroxo complex. The maximum rate of self-exchange reaction for the specific Cr^{3+}/Cr^{2+} couple is 2×10^5 M^{-1} s^{-1} however for the Ru^{3+}/Ru^{2+} redox couple it has not been directly measured. Although, rationally the specific rate for Ru^{3+}/Ru^{2+} redox couple can be considered close to that for $[Ru(NH_3)_6]^{3+}/[Ru(NH_3)_6]^{2+}$ that is 1×10^3 M^{-1} s^{-1}.

The mechanism of outer sphere redox reactions initiates with the formation of precursor complex which further undergo rearrangement route to become suitable for electron transfer process.

In these reactions, one of the reacting species utilizes the second coordination sphere or outer sphere of other reacting species. Both the reactants involved must be inert towards substitution to prefer outer sphere mechanism over that of inner sphere mechanism.

General mechanism:

$$[Ox]+[Red] \xrightleftharpoons{K_A} \{Ox\|Red\}$$

$$\{Ox\|Red\} \xrightleftharpoons{K_{ET}} \{Ox^-\|Red^+\}$$

$$\{Ox^-\|Red^+\} \xrightleftharpoons{K_{diss}} [Ox^-]+[Red^+]$$

$$K_{OBS} = K_A K_{EL}$$

In the complex form, it can be depicted as:

1. Precursor complex formation is basically that where both the reactants are weakly interacting to each other without changing their oxidation state and indeed in equilibrium.

$$M^{(a+1)+}L_x + M^{b+}L_y \xrightleftharpoons{K_{eq}} \left[M^{(a+1)+}L_x ----M^{b+}L_y\right] \text{Precursor complex}$$

2. Reorganization of the precursor complex to become suitable for electron transfer followed by relaxation of the successor complex.

$$M^{(a+1)+}L_x ----- M^{b+}L_y \xrightarrow{K_{ET}} \left[M^{a+}L_x ---- M^{(b+1)+}L_y\right] \text{Successor complex}$$

3. Dissociation of the successor complex eventually leads to the formation of ultimate products where both the complexes have changed their oxidation state by one.

$$\left[M^{a+}L_x ----- M^{(b+1)+}L_y\right] \xrightarrow{Kdiss} \left[M^{a+}L_x\right] + \left[M^{(b+1)+}L_y\right]$$

The rate of reaction can be determined from step 2 which is the slowest step of the reaction whereas formation of precursor complex as well as dissociation of the successor complex is quite rapid.

The thermodynamic energy profile diagram for outer sphere reaction is given below:

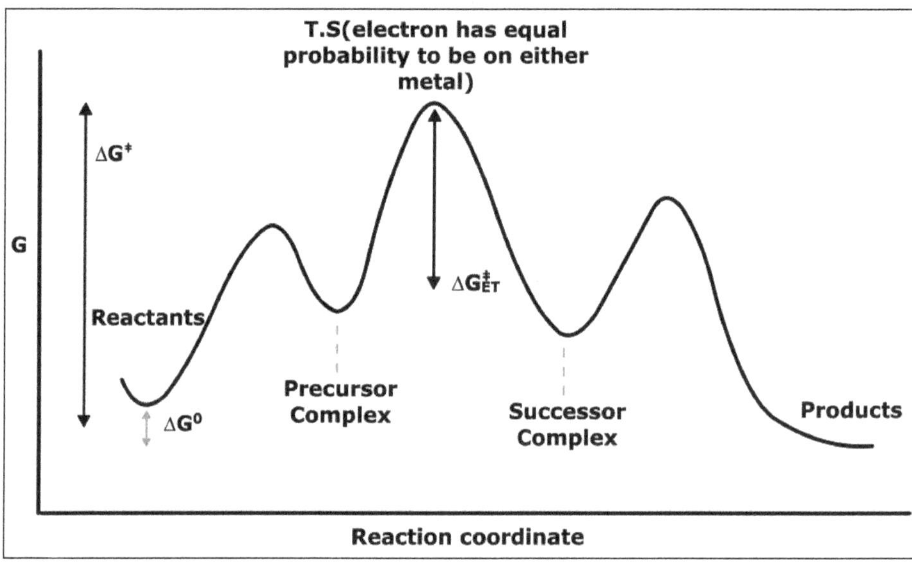

$\Delta G^\ddagger = \Delta Gt^\ddagger$ (energy required to bring reactants together) $+\Delta G_0^\ddagger$ (energy required for solvent recombination) $+\Delta G_i^\ddagger$ (energy needed for change in bond lengths by bringing orbitals in same energy level).

Mechanisms of Redox Reactions: Inner Sphere Reactions

The inner sphere reactions are those which incorporate "sharing/bridging" of at least one ligand for electron transfer and the reactants undergo change in complementary oxidation state. The reactants join together to result a single primary bond system through adopting ligand bridging pathway. There are two significant points for inner sphere reactions to take place; 1) The complex having at least one ligand with secondary binding site that is sterically favorable to the second metal ion is desirable (not compulsory) condition for inner sphere reactions; 2) Complex receiving the ligand must be labile in terms of expanding its coordination number at the time of electron transfer. Therefore, such reactions can be considered to be mediated by bridging ligand. Unlike to that for outer sphere mechanism, the coordination sphere of the complex participates in the inner sphere mechanism and atom/group/ion transfer occurs.

Despite satisfying abovementioned terms and conditions, the reactants may still adopt a lower energy pathway that is outer sphere mechanism therefore a convincing evidence for inner sphere route is required. Although, the variation in mechanistic plausibility has been numerously argued to support inner sphere pathway but direct evidences are still lacking. The argument came to end after the detailed and evident experiments performed by Taube and Myers employing several ligands that were obviously detected in the product. Indeed, the bridging/shared ligand is not initially connected with coordination sphere of metal centre. Eventually, the inertness of the product governs the inner sphere redox reaction timescale. The inert sphere reactions are of considerable importance owing to its widespread applications. For instance, beside universal use of Cr(II) as reductant, $[Fe(H_2O)_6]^{2+}$, $[Co(CN)_6]^{3-}$, $[V(H_2O)_6]^{2+}$ and $[Ru(NH_3)5(H_2O)_6]^{2+}$ etc. are being utilized as reductants.

The redox process can plausibly be accomplished through various routes for polyatomic bridging ligands while incoming metal binds with just one of the diverse bridging ligands available. Reaction between $[(NH_3)_5Co(\text{-}NC_6H_4CONH_2)]^{3+}$ and $[Cr(H_2O)_6]^{2+}$ possesses two probable remote binding sites. The activated complex has been structurally authenticated that incorporates bonding of $[Cr(H_2O)_6]^{2+}$ with the remote carbonyl moiety. A major deliberation deals with the atomic geometry upon chelation with the activated complex in due course of reaction.

The general representation of inner sphere reactions as follows:

$$\left[M^{III}L_5X\right]^{n+} + \left[M^{II}(Y)_6\right]^{2+} \xrightarrow{H^+} \left[M^{II}L_5\right]^{n+} + \left[M^{III}(Y)_6X\right]^{2+}$$

$$\downarrow$$

$$\left[M^{II}(Y)6\right]^{2+} + \left[M^{III}(Y)_5X\right]^{2+}$$

$$L = \text{Ligand}\left(\text{generally } NH_3\right); X = \text{bridging ligand; } Y = \text{Ligand}\left(\text{generally } H_2O\right)$$

A representative reaction is given below:

$$\left[Co\left(NH_3\right)_5{}^{*}CI\right]^{2+} + \left[Cr\left(H_2O\right)_6\right]^{2+} + 5\left[H_3O\right]^{+}$$

$$\downarrow$$

$$\left[Co\left(H_2O\right)_6\right]^{2+} + \left[Cr\left(H_2O\right)_5{}^{*}CI\right]^{2+} 5\left[NH_4\right]^{+}$$

The above reaction which involves inner sphere path proceeds with appreciably high rate (k = 6 × 10^5 M^{-1} sec^{-1}). However if we replace the Cl by NH_3 from Co(III) complex such as:

$$[Co(NH\ )_6] \quad [Cr(H\ O)_6] \quad 6[H\ O]$$
$$\downarrow$$
$$[Co(H_2O)_6]^{2+} \quad [Cr(H_2O)_6]^{3+} \quad 6[NH\ ]^{+}$$

This reaction proceeds through outer sphere path with considerable slower rate (k = 10^{-3} M^{-1} sec^{-1}).

The comparative inner sphere and outer sphere reaction pathways strongly recommend the presence of a bridging ligand in one of the metal complex reactants to precede the reaction through inner sphere pathway. The bridging ligands are usually negatively charges monodentate ligands i.e.

Cl^-, Br^-, I^-, OH^-, NCS^- or SCN^-, $N3^-$, CN^- etc.

There are some other ligands as well which can act as bridging ligands in the course of electron transfer reaction through inner sphere path such as:

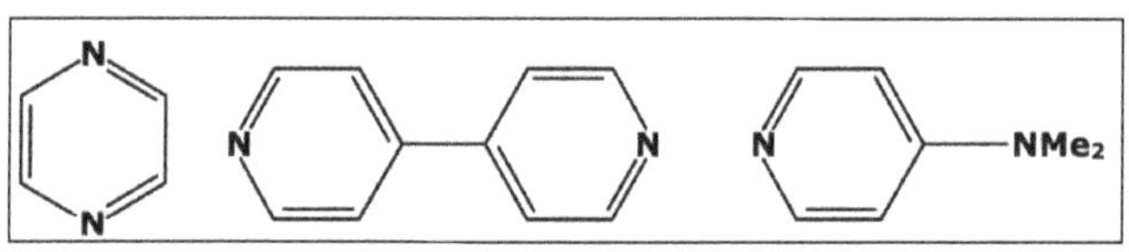

General considerations of inner sphere redox reaction mechanism are:

These reactions usually proceed in three significant steps:

- Both the metal complexes should connect with a bridging ligand during the course of substitution. One of the reacting species should have a ligand that can form a bridge between both the metal complexes. The bridging ligand formation is basically intermediate therefore it is occasionally called a ligand-bridged mechanism. For instance, H_2O and NH_3 are monodentate ligands but these cannot serve as bridging ligands, whereas Cl– and CN– can act as bridging ligand.
- The transfer of electron as well as sometimes transfer of ligand occurs depending upon the groups or ligands attached with the reactants. However, inner-sphere redox reactions are slower relative to that of outer-sphere reactions.
- Eventually, the products are separated to form inert complexes.

The common reaction mechanism is:

$$L_5M^{III}X + M'^{II}Y_6 \underset{K_{-a}}{\overset{K_a}{\rightleftharpoons}} \{L_5M^{III} - X - M'^{II}Y_5\} + Y$$

$$\{L_5M^{III}-X-M'^{II}Y_5\}\xrightarrow{K_{ET}}\{L_5M^{II}-X-M'^{III}Y_5\}$$
$$Y+\{L_5M^{II}-X-M'^{III}Y_5\}\xrightarrow{K_d}L_5M^{II}Y+XM'^{III}Y_5$$

Or

$$Y+\{L_5M^{II}-X-M'^{III}Y_5\}\xrightarrow{K_d}L_5M^{II}X+M'^{III}Y_6$$

The bridging/sharing ligand mediates inner sphere electron transfer process. It is evident from the above depicted general mechanism that:

- A ligand is shared/bridged between oxidant and reductant both in the precursor and successor complexes.
- The electron transfer takes place between both the metals only after activation through bridging of the ligand.
- The transfer of ligand may also occur between metal complexes however it is not a essential requirement.

The fundamentals for Inner Sphere Mechanism are:

- One of the reacting metal complexes (generally oxidant) has at least one ligand which can bridge concurrently to both the metals (in the oxidant and reductant).
- The metal complex (usually reductant) must be labile for substitution so that one of the ligand from it can be displaced by the incoming bridging ligand.
- Another schematic representation for the inner sphere mechanism has been given below which demonstrates basic steps along with rate expression:

$$Ox-X+Red(H_2O)\overset{K_1}{\rightleftharpoons}Ox---x---Red+H_2O$$

Precursor complex formation

$$Ox---x---Red\overset{K_2}{\rightleftharpoons}{}^{-}Ox---x---Red^{+}$$

Precursor activation followed by e^- transfer

$${}^{-}Ox---X---Red^{+}+H_2O\overset{K_3}{\rightleftharpoons}Ox(H_2O)^{-}+RedX^{+}$$

Separation of the Products

Therefore, overall reaction can be represented as:

$$Ox-x+Red\rightleftharpoons Ox-x-Red\rightleftharpoons Ox^{-}+Red^{+}$$

$$K_{obs} = \frac{K_1K_3}{K_2 + K_3}[Ox - X][Red]$$

Example and Mechanism

$$\left[Co(NH_3)_5{}^*CI\right]^{2+} + \left[Cr(H_2O)_6\right]^{2+} + 5\left[H_3O\right]^{+}$$

Co(III)LS, Intert, d^6 ; Cr(III), Labile, d^4

$$\downarrow$$

$$\left[Co(H_2O)_6\right]^{2+} + \left[Cr(H_2O)_5{}^*CI\right]^{2+} + 5\left[NH_4\right]^{+}$$

Co(II)HS, Labile, d^7 ; Cr(III), Inert, d^3

The entire Cr(III) complex $[Cr(H_2O)5{}^*Cl]^{2+}$ formed witnessed the presence of isotopic *Cl (*Cl = ^{36}Cl which clearly suggests that the source of Cl ligand is oxidant complex $[Co(NH_3)_6{}^*Cl]^{2+}$ and it is same ligand that bridges both the metal centers.

Mechanism:

$$\left[Co(NH_3)_5{}^*CI\right]^{2+} + \left[Cr(H_2O)_6\right]^{2+} \underset{K_{-1}}{\overset{K_1}{\rightleftharpoons}} \left[(NH_3)_5CO^{III}{}_-{}^*CI - Cr^{II}(H_2O)_5\right]^{4+} + H_2O(fast)$$

$$\left[(NH_3)_5Co^{III} - CI - Cr^{II}(H_2O)_5\right]^{4+} \underset{K_{-2}}{\overset{K_2}{\rightleftharpoons}} \left[(NH_3)_5CO^{II} -{}^* CI - Cr^{III}(H_2O)_5\right]^{4+}$$

$$\left(e^- \text{ transfer, slow}\right)$$

$$\left[(NH_3)_5CO^{II} - CI - -Cr^{III}(H_2O)_5\right]^{4+} \underset{K_{-3}}{\overset{K_3}{\rightleftharpoons}} \left[CO^{II}(NH_3)_5(H_2O)\right]^{2+} \left[Cr^{II}(H_2O)_5{}^*CI\right]^{2+}$$

The slowest step is the electron transfer process which is the rate determining step. However, in general, rate determining step may be electron transfer step or bridgebreaking step or bridge-forming step. Majority of the inner-sphere redox reactions exhibits second order kinetics. The energetic of above mentioned reaction is given.

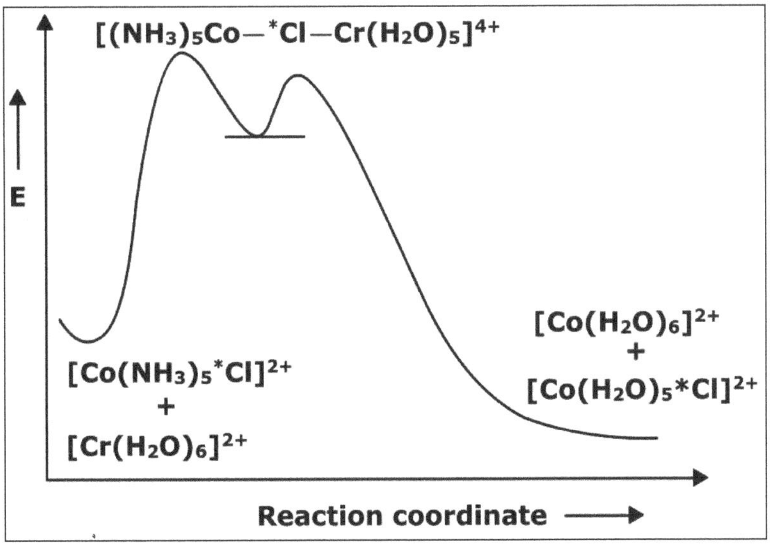

$$\left[Co(NH_3)_5{}^*Br\right]^{2+} + \left[V(H_2O)_6\right]^{2+} + 5\left[H_3O\right]^{+}$$
$$\downarrow$$
$$\left[Co(H_2O)_6\right]^{2+} + \left[V(H_2O)_5{}^*Br\right]^{2+} + 5\left[NH_4\right]^{+}$$

Mechanism:

$$\left[Co(NH_3)_5{}^*Br\right]^{2+} + \left[Cr(H_2O)_6\right]^{2+} \rightarrow \left[(NH_3)_5Co^{III}-{}^*Br-V^{II}(H_2O)_6\right]^{4+} \text{(fast)}$$
$$\left[(NH_3)_5Co^{III}-{}^*Br-V^{II}(H_2O)_5\right]^{4+} \rightarrow \left[(NH_3)_5Co^{III}\right]^{2+}-{}^*Br-\left[V^{III}(H_2O)_6\right]^{3+} \text{(fast)}$$
$$\left[(NH_3)_5Co^{II}\right]^{2+}-{}^*Br-\left[V^{III}(H_2O)_6\right]^{3+} \rightarrow \left[Co^{II}(NH_3)_5\right]^{2+} + \left[V^{III}(H_2O)_5{}^*Br\right]^{2+} + H_2O\text{(slow)}$$

Relative studies on above reactions, it is obvious that bridging group imposes subtle influence on rate constant (k) and the rate is analogous to that for H_2O exchange. The slowest step is dissociation of H_2O which is also rate determining step, not bridge forming step or electron transfer step.

Metals and Alloys

In the chemistry of molecular compounds, we are accustomed to the idea that properties depend strongly on structure. For example we can rationalize the polarity of the water molecule based on its shape. We also know that two molecules with the same composition (e.g., ethanol and dimethyl ether) have very different properties based on the bonding arrangements of atoms. It should come as no surprise that the properties of extended solids are also connected to their structures, and so to understand what they do we should begin with their crystal structures. Most of the metals in the periodic table have relatively simple structures and so this is a good place to begin.

Unit Cells and Crystal Structures

Possible unit cells in a periodic tile pattern. The arrows connect translationally equivalent points (lattice points) in the pattern.

Crystals can be thought of as repeating patterns, much like wallpaper or bathroom tiles, but in three dimensions. The fundamental repeating unit of the crystal is called the unit cell. It is a three dimensional shape that can be repeated over and over by unit translations to fill space (and leave no gaps) in the structure. Some possible unit cells are shown in the tiling pattern at the right, along with arrows that indicate unit translation vectors. In three dimensions, the hexagonal or rhombic unit cells of this pattern would be replaced by three dimensional boxes that would stack together to fill all space. As shown in the figure, the origin of the unit cell is arbitrary. The same set of boxes will fill all space no matter where we define the origin of the lattice. We will see that pure metals typically have very simple crystal structures with cubic or hexagonal unit cells. However the crystal structures of alloys can be quite complicated.

When considering the crystal structures of metals and alloys, it is not sufficient to think of each atom and its neighboring ligands as an isolated system. Instead, think of the entire metallic crystal as a network of atoms connected by a sea of shared valence electrons. The electrons are delocalized because there are not enough of them to fill each "bond" between atoms with an electron pair. For example, in the crystal structures of s-block and p-block metals, each atom has either 8 or 12 nearest neighbors, but the maximum number of s + p electrons is 8. Thus, there are not enough to put two electrons between each pair of atoms. Transition metals can also use their d-orbitals in bonding, but again there are never enough electrons to completely fill all the "bonds."

The atoms in a metal lattice arrange themselves in a certain pattern which can be represented as a 3D box structure known as the unit cell which repeats across the entire metal.

Simple Cubic	Body Centered Cubic	Face Centered Cubic	Hexagonal Close Packed
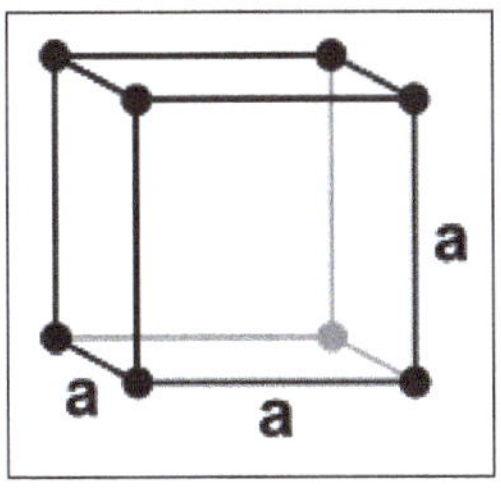	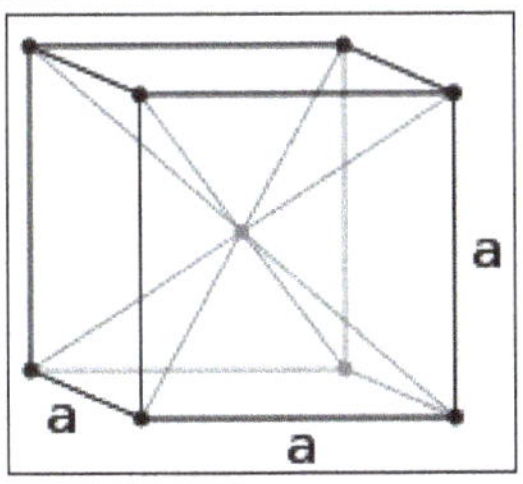	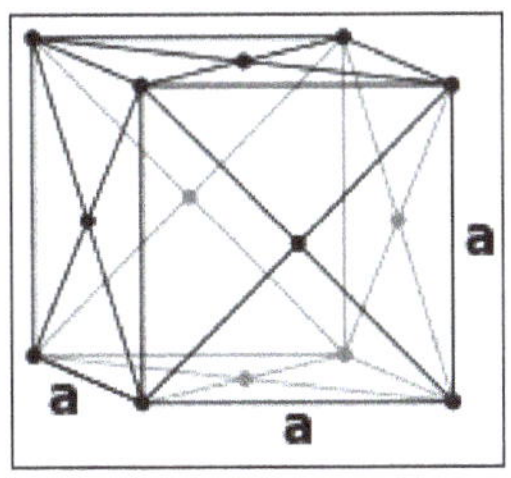	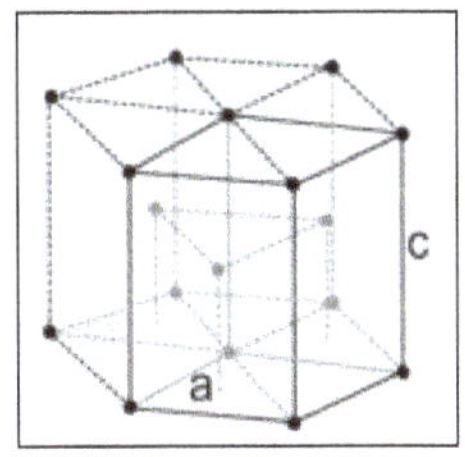
1 atom/cell	2 atoms/cell	4 atoms/cell	2 atoms/cell

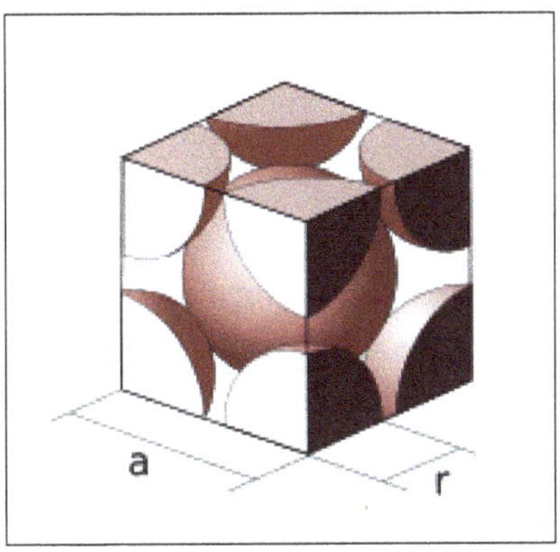

The unit cell of a bcc metal contains two atoms.

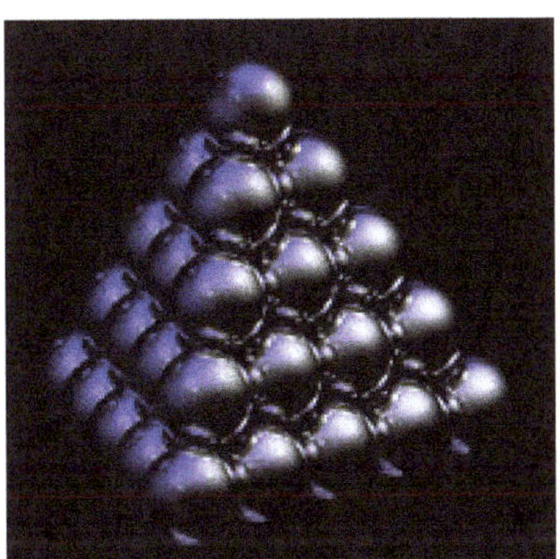

Face-centered cubic stack of cannonballs.

Metal atoms can be approximated as spheres, and therefore are not 100 % efficient in packing, the same way a stack of cannonballs has some empty spaces between the balls.

Different unit cells have different packing efficiencies. The number of atoms that is included in the unit cell only includes the fractions of atoms *inside* of the box. Atoms on the corners of the unit cell count as ⅛ of an atom, atoms on a face count as ½, an atom in the center counts as a full atom. Using this, let's calculate the number of atoms in a simple cubic unit cell, a face centered cubic (fcc) unit cell, and a body centered cubic (bcc) unit cell.

- Simple Cubic: 8 corner atoms × ⅛ = 1 atom/cell. The packing in this structure is not efficient (52%) and so this structure type is very rare for metals.
- Body Centered Cubic, bcc: (8 corner atoms × ⅛) + (1 center atom × 1)= 2 atoms/cell. The packing is more efficient (68%) and the structure is a common one for alkali metals and early transition metals. Alloys such as brass (CuZn) also adopt these structures.
- Face Centered Cubic, fcc (also called Cubic Close Packed, ccp): (8 corner atoms × ⅛)+ (6 face atoms × ½)= 4 atoms/cell. This structure, along with its hexagonal relative (hcp), has the most efficient packing (74%). Many metals adopt either the fcc or hcp structure.
- Hexagonal Close Packed, hcp: Like the fcc structure, the packing density of hcp is 74%.

Calculating the Packing Fraction

The packing fractions of the crystal structures shown above can be calculated by dividing the volume of the atoms in the cell by the volume of the cell itself. The volume of the atoms in the cell is equal to the number of atoms/cell times the volume of a sphere, $(4/3)\pi r^3$. The volume of the cubic cells is found by cubing the side length. As an example, let's calculate the packing efficiency of a simple cubic unit cell. a simple cubic unit cell contains one atom. The side length of the simple cubic unit cell is 2r, since the centers of each atom occupy the corners of the unit cell.

Packing efficiency=$[(1 \text{ atom})\times(4/3)\pi r^3]/[(2r)^3]=0.523$.

The same method can be applied to bcc and fcc structures.

Bravais Lattices

Crystal lattices can be classified by their translational and rotational symmetry. In three-dimensional crystals, these symmetry operations yield 14 distinct lattice types which are called Bravais lattices. In these lattice diagrams (shown below) the dots represent lattice points, which are places where the whole structure repeats by translation. For example, in the body-centered cubic (bcc) structure of sodium metal, which is discussed below, we put one atom at the corner lattice points and another in the center of the unit cell. In the NaCl structure, we place one NaCl formula unit on each lattice point in the face-centered cubic (fcc) lattice. That is, one atom (Na or Cl) would be placed on the lattice point and the other one would be placed halfway between. Similarly, in the cubic diamond structure, we place one C_2 unit around each lattice point in the fcc lattice.

The fourteen Bravais lattices fall into seven crystal systems that are defined by their rotational symmetry. In the lowest symmetry system (triclinic), there is no rotational symmetry. This results in a unit cell in which none of the edges are constrained to have equal lengths, and none of the angles are 90°. In the monoclinic system, there is one two-fold rotation axis (by convention, the b-axis), which constrains two of the angles to be 90°. In the orthorhombic system, there are three mutually perpendicular two-fold axes along the three unit cell directions. Orthorhombic unit cells have three unequal unit cell edges that are mutually perpendicular. Tetragonal unit cells have a four-fold rotation axis which constrains all the angles to be 90° and makes the a and b axes equivalent. The rhombohedral system has a three-fold axis, which constrains all the unit cell edges and angles to be equal, and the hexagonal system has a six-fold axis, which constrains the a and b lattice dimensions to be equal and the angle between them to be 120°. The cubic system has a three-fold axis along the body diagonal of the cube, as well as two-fold axes along the three perpendicular unit cell directions. In the cubic system, all unit cell edges are equal and the angles between them are 90°.

The translational symmetry of the Bravais lattices (the lattice centerings) are classified as follows:

- Primitive (P): Lattice points on the cell corners only (sometimes called simple).
- Body-Centered (I): Lattice points on the cell corners with one additional point at the center of the cell.
- Face-Centered (F): Lattice points on the cell corners with one additional point at the center of each of the faces of the cell.
- Base-Centered (A, B, or C): Lattice points on the cell corners with one additional point at the center of each face of one pair of parallel faces of the cell (sometimes called end-centered).

Not all combinations of the crystal systems and lattice centerings are unique. There are in total 7 × 6 = 42 combinations, but it can be shown that several of these are in fact equivalent to each other. For example, the monoclinic I lattice can be described by a monoclinic C lattice by different choice of crystal axes. Similarly, all A- or B-centred lattices can be described either by a C- or P-centering. This reduces the number of combinations to 14 conventional Bravais lattices, shown in the table below.

When the fourteen Bravais lattices are combined with the 32 crystallographic point groups, we obtain the 230 space groups. These space groups describe all the combinations of symmetry operations that can exist in unit cells in three dimensions. For two-dimensional lattices there are only 17 possible plane groups, which are also known as wallpaper groups.

Crystal family	Lattice system	Schönflies	14 Bravais Lattices			
			Primitive	Base-centered	Body-centered	Face-centered
Triclinic		C_i	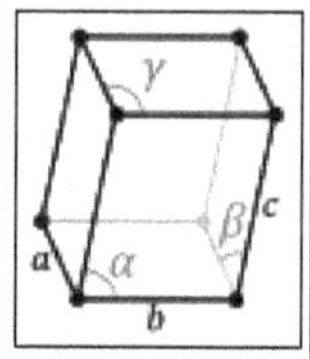			
Monoclinic		C_{2h}	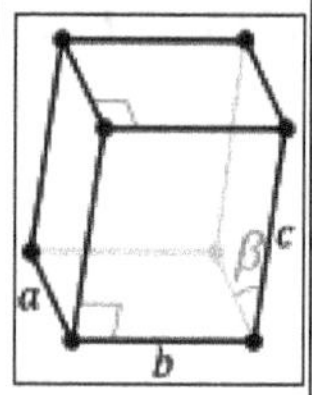	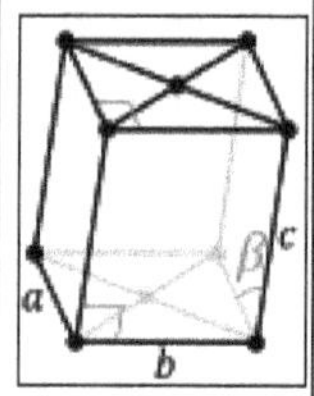		
Orthorhombic		D_{2h}	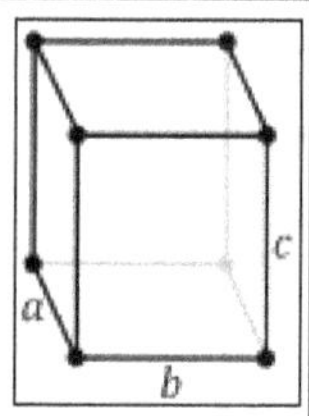	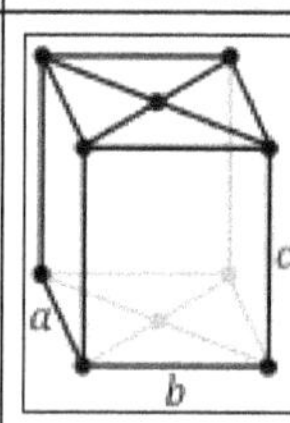	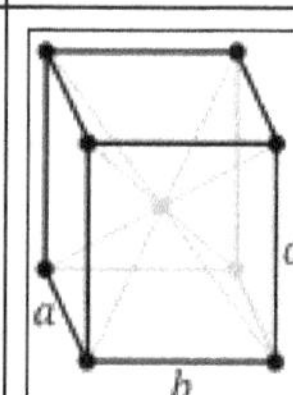	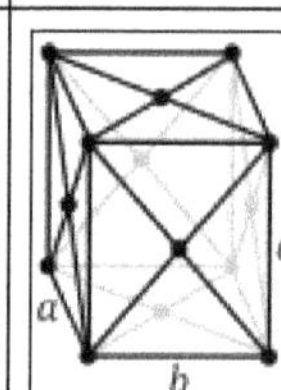
Tetragonal		D_{4h}	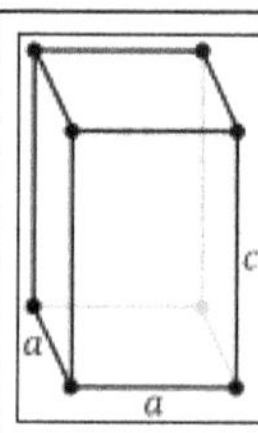		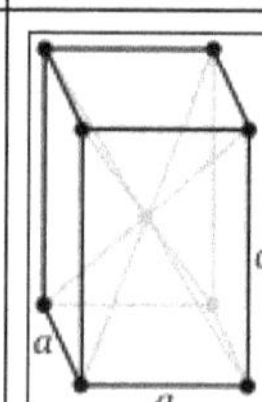	
Hexagonal	Rhombohedral	D_{3d}	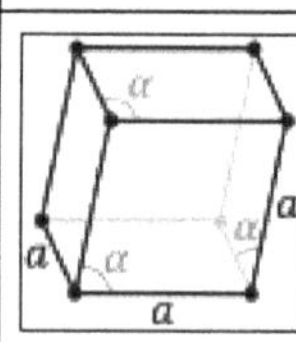			
	Hexagonal	D_{6h}	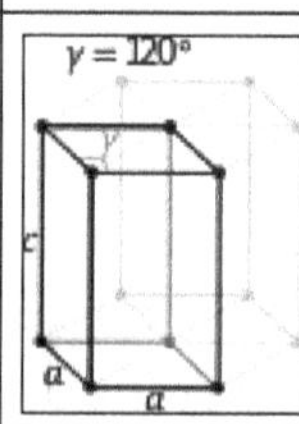			
Cubic		O_h	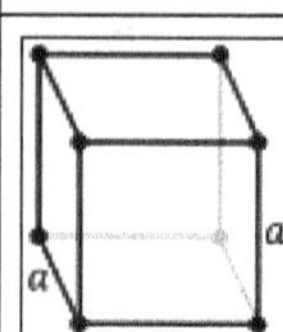		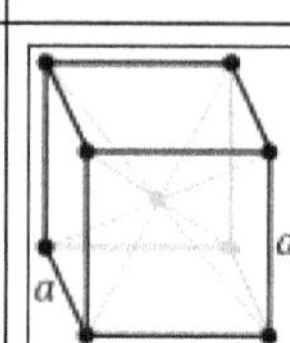	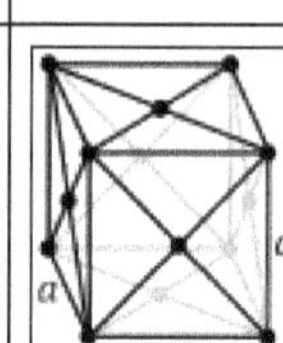

Crystal Structures of Metals

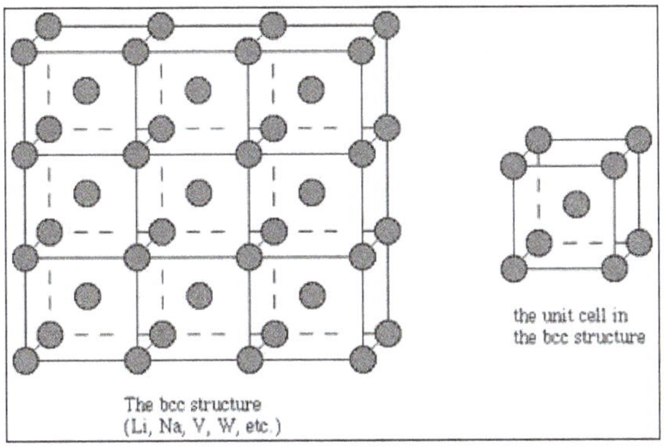

Body-centered cubic.

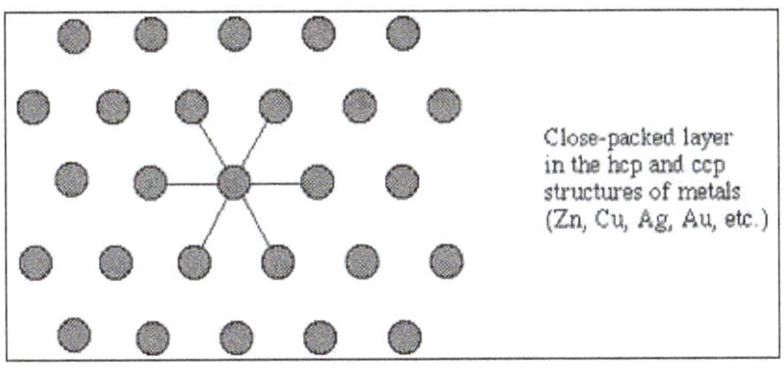

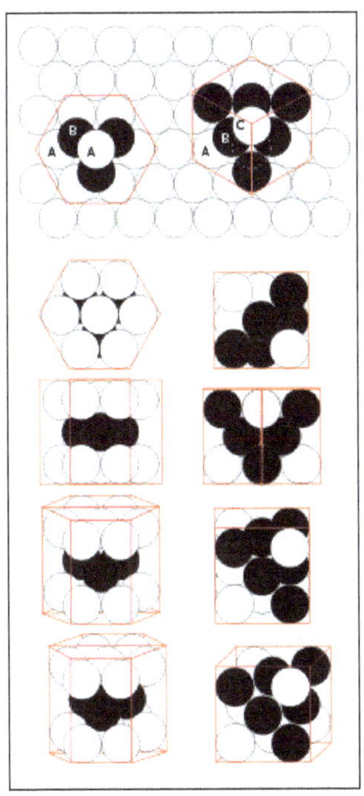

hcp (left) and fcc (right) close-packing of spheres.

The Crystalline Nature of Metals

All metallic elements (except Cs, Ga, and Hg) are crystalline solids at room temperature.

Like ionic solids, metals and alloys have a very strong tendency to crystallize, whether they are made by thermal processing or by other techniques such as solution reduction or electroplating. Metals crystallize readily and it is difficult to form a glassy metal even with very rapid cooling. Molten metals have low viscosity, and the identical (essentially spherical) atoms can pack into a crystal very easily. Glassy metals can be made, however, by rapidly cooling alloys, particularly if the constituent atoms have different sizes. The different atoms cannot pack in a simple unit cell, sometimes making crystallization slow enough to form a glass.

Crystal Structures

Most metals and alloys crystallize in one of three very common structures: body-centered cubic (bcc), hexagonal close packed (hcp), or cubic close packed (ccp, also called face centered cubic, fcc). In all three structures the coordination number of the metal atoms (i.e., the number of equidistant nearest neighbors) is rather high: 8 for bcc, and 12 for hcp and ccp. We can contrast this with the low coordination numbers (i.e., low valences - like 2 for O, 3 for N, or 4 for C) found in nonmetals. In the bcc structure, the nearest neighbors are at the corners of a cube surrounding the metal atom in the center. In the hcp and ccp structures, the atoms pack like stacked cannonballs or billiard balls, in layers with a six-coordinate arrangement. Each atom also has six more nearest neighbors from layers above and below. The stacking sequence is ABCABC... in the ccp lattice and ABAB... in hcp. In both cases, it can be shown that the spheres fill 74% of the volume of the lattice. This is the highest volume fraction that can be filled with a lattice of equal spheres.

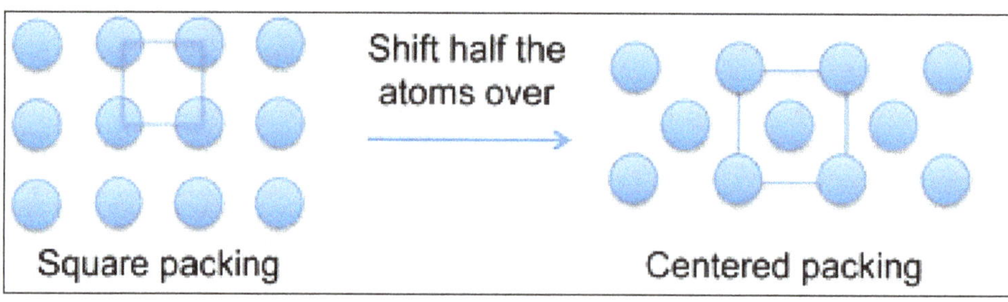

Atoms in metallic crystals have a tendency to pack in dense arrangments that fill space efficiently. The simple square packing (left) upon which the simple cubic structure is based is inefficient and thus rare among metallic crystal structures. Body- or face-centered structures fill space more efficiently and more common.

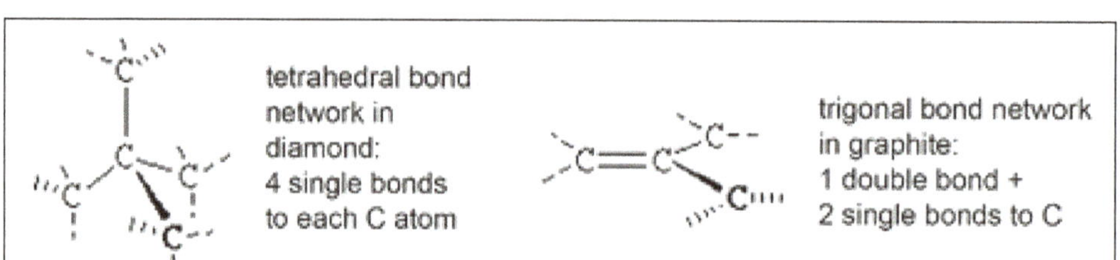

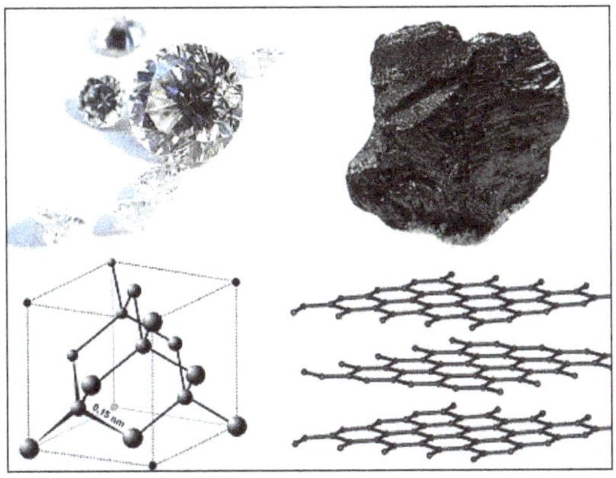

Periodic Trends in Structure and Metallic Behaviour

Remember where we find the metallic elements in the periodic table - everywhere except the upper right corner. This means that as we go down a group in the p-block (let's say, group IVA, the carbon group, or group VA, the nitrogen group), the properties of the elements gradually change from nonmetals to metalloids to metals. The carbon group nicely illustrates the transition. Starting at the top, the element carbon has two stable allotropes - graphite and diamond. In each one, the valence of carbon atoms is exactly satisfied by making four electron pair bonds to neighboring atoms. In graphite, each carbon has three nearest neighbors, and so there are two single bonds and one double bond. In diamond, there are four nearest neighbors situated at the vertices of a tetrahedron, and so there is a single bond to each one.

The two elements right under carbon (silicon and germanium) in the periodic table also have the diamond structure (recall that these elements cannot make double bonds to themselves easily, so there is no graphite allotrope for Si or Ge). While diamond is a good insulator, both silicon and germanium are semiconductors (i.e., metalloids). Mechanically, they are hard like diamond. Like carbon, each atom of Si and Ge satisfies its valence of four by making single bonds to four nearest neighbors.

The next element under germanium is tin (Sn). Tin has two allotropes, one with the diamond structure, and one with a slightly distorted bcc structure. The latter has metallic properties (metallic luster, malleability), and conductivity about 10^9 times higher than Si. Finally, lead (Pb), the element under Sn, has the ccp structure, and also is metallic. Note the trends in coordination number and conducting properties:

Element	Structure	Coord. no.	Conductivity
C	graphite, diamond	3, 4	semimetal, insulator
Si	diamond	4	semiconductor
Ge	diamond	4	semiconductor
Sn	diamond, distorted bcc	4, 8	semiconductor, metal
Pb	ccp	12	metal

The elements C, Si, and Ge obey the octet rule, and we can easily identify the electron pair bonds in their structures. Sn and Pb, on the other hand, adopt structures with high coordination numbers. They *do not have enough valence electrons* to make electron pair bonds to each neighbor (this is a common feature of metals). What happens in this case is that the valence electrons become "smeared out" or delocalized over all the atoms in the crystal. It is best to think of the bonding in metals as a crystalline arrangement of positively charged cores with a "sea" of shared valence electrons gluing the structure together. Because the electrons are not localized in any particular bond between atoms, they can move in an electric field, which is why metals conduct electricity well. Another way to describe the bonding in metals is nondirectional. That is, an atom's nearest neighbors surround it in every direction, rather than in a few particular directions (like at the corners of a tetrahedron, as we found for diamond). Nonmetals (insulators and semiconductors), on the other hand, have directional bonding. Because the bonding is non-directional and coordination numbers are high, it is relatively easy to deform the coordination sphere (i.e., break or stretch bonds) than it is in the case of a nonmetal. This is why elements like Pb are much more malleable than C, Si, or Ge.

Bonding in Metals

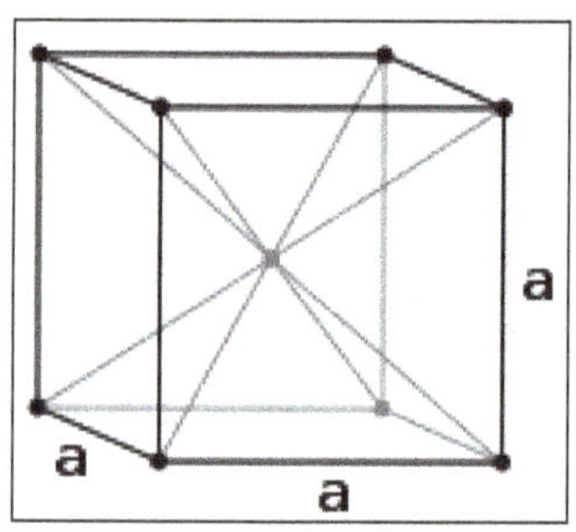

Sodium metal crystallizes in the body-centered cubic structure, in which each atom has eight nearest neighbors. Since the electronic configuration of Na is [Ar]$3s^1$, there are only two valence electrons per unit cell that are shared among eight Na-Na bonds. This means that the Na-Na bond order is 1/8 in Na metal.

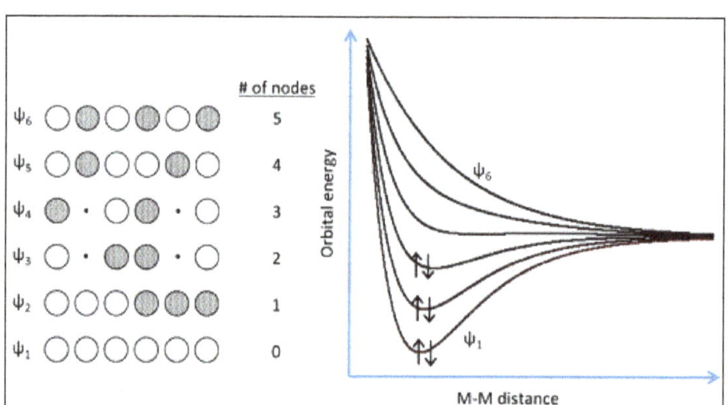

MO picture for a linear chain of six Na atoms.

Three of the six MOs can accommodate all six valence electrons. Adding more atoms to the chain makes more molecular orbitals of intermediate energy, which eventually merge into a continuous band of orbitals. For Na, the 3s band is always half-filled because each MO can accommodate two electrons.

The electron pair description of chemical bonds, which was the basis of the octet rule for p-block compounds, breaks down for metals. This is illustrated well by Na metal, the structure of which is shown at the left. Na has too few valence electrons to make electron pair bonds between each pair of atoms. We could think of the Na unit cell as having eight no-bond resonance structures in which only one Na-Na bond per cell contains a pair of electrons.

A more realistic way to describe the bonding in metals is through band theory. The evolution of energy bands in solids from simple MO theory is illustrated at the right for a chain of six Na atoms, each of which has one 3s valence orbital and contributes one valence electron. In general, n atomic orbitals (in this case the six Na 3s orbitals) will generate n molecular orbitals with n-1 possible nodes. the energy versus internuclear distance graph for a two hydrogen atom system has a low energy level and a high energy level corresponding to the bonding and antibonding molecular orbitals, respectively. These two energy levels were well separated from each other, and the two electrons in H_2 energetically prefer the lower energy level. If more atoms are introduced to the system, there will be a number of additional levels between the lowest and highest energy levels.

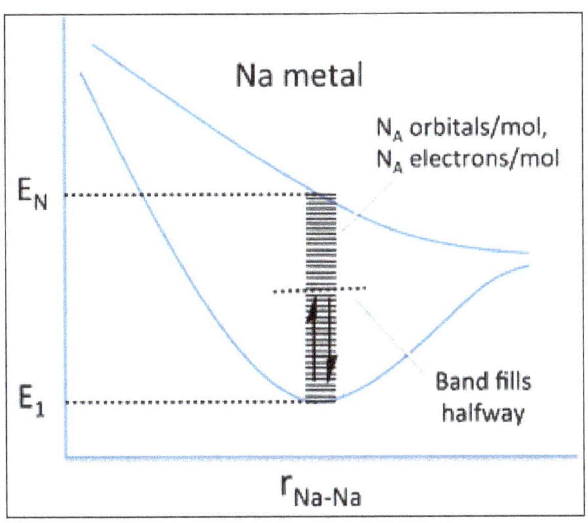

In band theory, the atom chain is extrapolated to a very large number - on the order of 10^{22} atoms in a crystal - so that the different combinations of bonding and anti-bonding orbitals create “bands” of possible energy states for the metal. In the language of physics, this approach of building the bands from discrete atomic orbitals is called the “tight-binding” approximation. The number of atoms is so large that the energies can be thought of as a continuum rather than a series of distinct levels. A metal will only partially fill this band, as there are fewer valence electrons than there are energy states to fill. In the case of Na metal this results in a half-filled 3s band.

- Nearly free electron model. In metals, the valence electrons are delocalized over many atoms. The total energy of each electron is given by the sum of its kinetic and potential energy:

  $E = KE + PE$

  $E \approx p^2/2m + V$

Where p is the momentum of the electron (a vector quantity), m is its mass, and V is an average potential that the electron feels from the positive cores of the atoms. This potential holds the valence electrons in the crystal but, in the free electron model, is essentially uniform across the crystal.

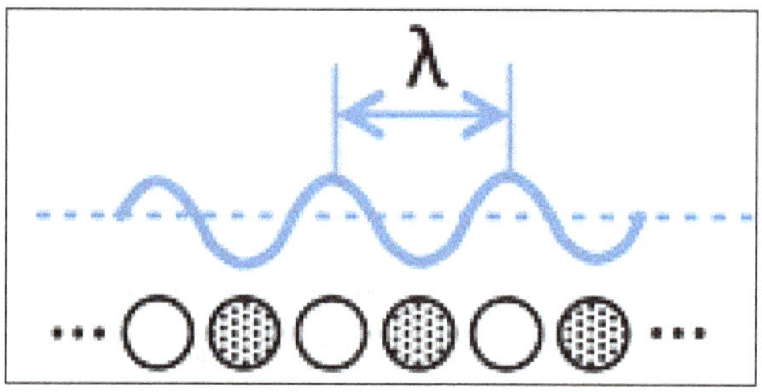

We can think of the wavelength of an electron in a molecular orbital as twice the distance between nodes. If there are N atoms in a linear chain, the wavelength of the nth orbital is given by $\lambda = 2Na/n$, where a is the distance between atoms.

Electron Wavelength and Wavenumber

What are the consequences of this model for band theory? For a hypothetical infinite chain (i.e., a 1D crystal) of Na atoms, the molecular orbitals at the bottom of the 3s band are fully bonding and the *wavelength* of electrons (2x the distance between nodes) in these orbitals is very long. At the top of the band, the highest orbital is fully antibonding and the wavelength is 2 times the distance between atoms (2a), since there is one node per atom. Remember that the wavelength of an electron (λ) is inversely proportional to its momentum p, according to the de Broglie relation $\lambda = h/p$.

For a (nearly) free electron, the kinetic energy can be expressed in terms of its wavelength, using $KE = p^2/2m$ and the de Broglie relation:

$$KE = p^2/2m = h^2/(2m\lambda^2)$$

At this point, it is convenient to define the wavenumber of the electron as k, which has units of inverse length and is inversely proportional to λ. k is also directly proportional to the momentum p. Like p, k is a vector quantity. In a 1D crystal, k can be either positive or negative, corresponding to an electron moving to the left or right along the chain.

$k = 2\pi/\lambda = 2\pi p/h$, where h is Planck's constant

The most important property of k is that it is *directly proportional to the number of nodes* n in a molecular orbital within the band. For a 1D crystal of sodium atoms that contains N unit cells, each separated by a distance a, a molecular orbital with n nodes has a wavelength $\lambda = 2Na/n$ and the wavenumber $k = \pi n/Na$. We can see from this definition that $k = 0$ at the bottom of the band (where λ is infinite) and $k = \pi/a$ at the top of the band where the MO contains N nodes and $\lambda = 2a$.

Energies of Orbitals in a Metallic Crystal

Electrons with long wavelengths do not "feel" the individual atoms in the lattice and so they behave as if they are *nearly free* (but confined to the crystal). Near the bottom of the band, the electron energy increases *parabolically* with the number of nodes ($KE \propto n^2$), since the momentum p is directly proportional to n. Because p is also directly proportional to k, we can write:

$$KE = p^2/2m = h^2k^2/8\pi^2m$$

This parabolic relationship is followed as long as the electron wavelength is long compared to the distance between atoms. Near the top of the band, the wavelength becomes shorter and the electrons start to feel the positively charged atomic cores. In particular, the electrons prefer to have the maxima in their wave functions line up with the atomic cores, which is the most electrostatically favorable situation. The electron-atom attraction lowers the energy and causes the E vs. k curve to deviate from the parabolic behavior of a "free" electron as shown in the figure below.

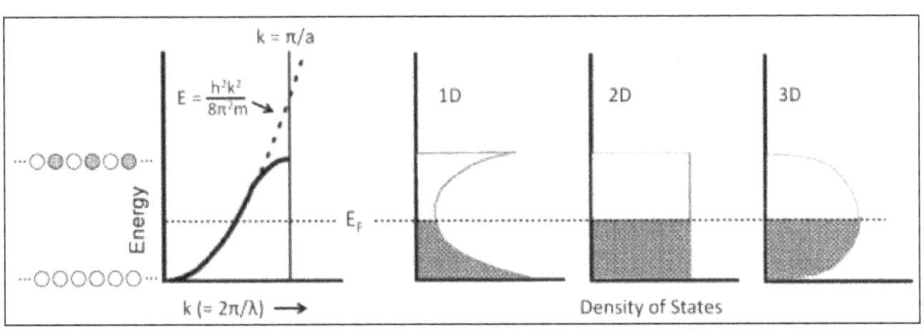

Electrons in metals follow a parabolic dispersion curve, where the energy increases with the square of the wavenumber, k. Near the top of the band, the dispersion curve deviates from the parabolic dotted line. Because there is one MO for each value of k, the number of orbitals per unit energy (the density of states, DOS) is highest at the bottom and top of the band for a 1D chain of atoms. The density of states is constant with energy for a 2D crystal, and has a maximum in the middle of the band for a 3D crystal. At low temperature, all the MOs below the Fermi level E_F are occupied, and all the MOs above it are empty.

Density of States (DOS)

The density of states is defined as the number of orbitals per unit of energy within a band. Because of the parabolic relation between E and k, the density of states for a 1D metallic crystal is highest near the bottom and top of the energy band where the slope of the E vs. k curve is closest to zero. The shape of the DOS curve is different in crystals of higher dimensionality as shown in the figure, because statistically there are more ways to make an orbital with N/2 nodes than there are with zero or N nodes. The situation is analogous to the numbers you can make by rolling dice. With one die, the numbers 1-6 have equal probability. However, with two dice there is only one way to make a two (snake eyes) or a twelve (boxcars), but many ways to make a seven (a winner).

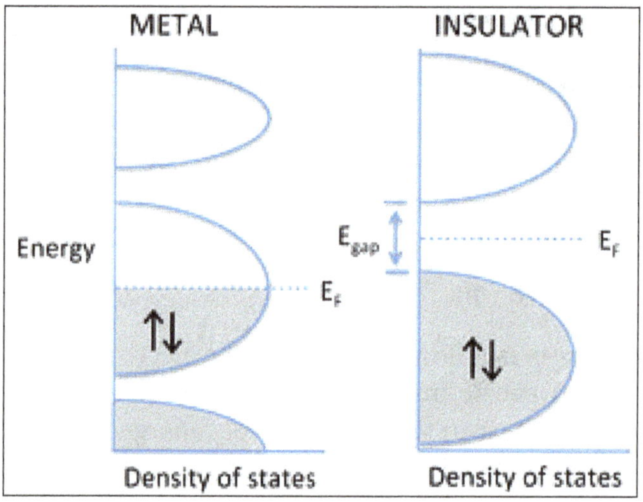

While most of the time we will talk about 3D crystals that have their maximum DOS near the middle of the energy band, there are examples of quasi-1D systems, such as carbon nanotubes. Metallic carbon nanotubes have strong optical absorption bands that correspond to transitions between the two regions of high DOS (the van Hove singularities) near the bottom and top of the bands.

Metals, Semiconductors and Insulators

The degree to which bands fill determines whether a crystalline solid is a metal, semiconductor, or insulator. If the highest occupied molecular orbitals lie within a band - i.e., if the Fermi level E_F cuts through a band of orbitals - then the electrons are free to change their speed and direction in an electric field and the solid is metallic. However, if the solid contains just enough electrons to completely fill a band, and the next highest set of molecular orbitals is empty, then it is a semiconductor or insulator. In this case, there is an energy gap between the filled and empty bands, which are called the valence and conduction bands, respectively. Although the distinction is somewhat arbitrary, materials with a large gap (> 3 eV) are called insulators, and those with smaller gaps are called semiconductors.

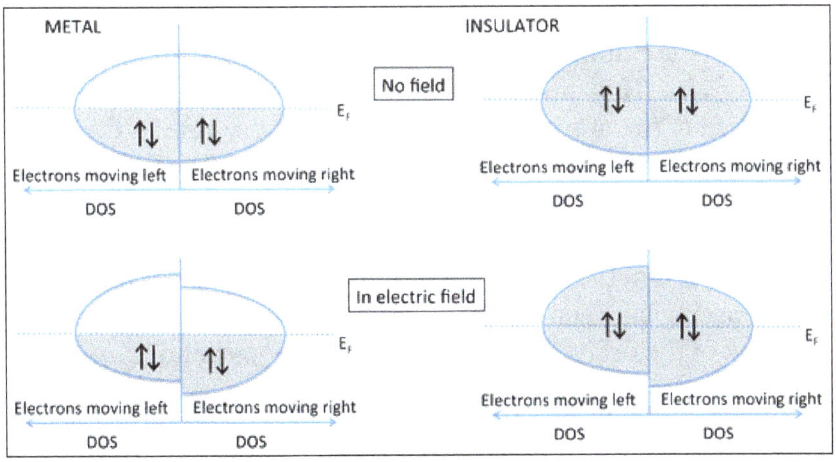

E vs. DOS diagrams comparing the behavior of metals and insulators in an applied electric field.

Why don't insulators conduct electricity? The energy vs. DOS diagram to the right shows what happens when an electric field is applied to a metal or an insulator. In this case we have changed the diagram to show explicitly the energies of electrons moving left and right. These energies are the same in the absence of an electric field. Once we apply a field (e.g., by putting a voltage across a metal wire), the electrons moving in the direction of the field have lower energy than those moving in the opposite direction. In the case of the metal, the populations of electrons moving with and against the electric field are different, and there is a net flow of current. Note that this can happen only when the Fermi level cuts through a partially filled band. With a semiconductor or insulator, the valence band is filled and the conduction band is empty. Applying an electric field changes the energies of electrons traveling with and against the field, but because the band is filled, the same number are going in both directions and there is no net current flow.

Note that in this picture, all the molecular orbitals extend over the entire crystal. The valence electrons are delocalized, even in the case of a semiconductor or insulator. However, there can be no net movement of electrons unless the band is partially filled.

Conduction in Metals

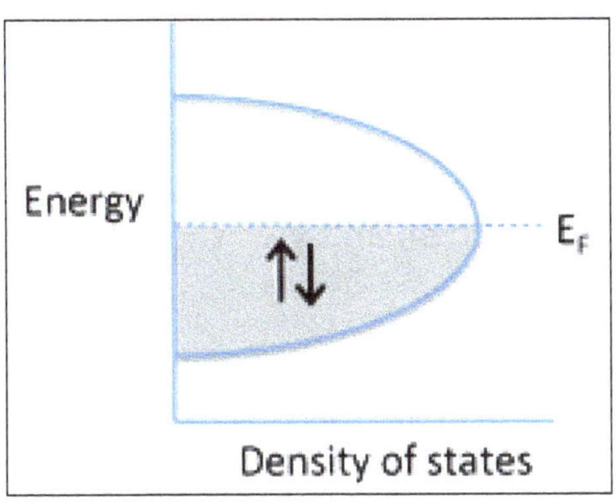

In metals, the valence electrons are in molecular orbitals that extend over the entire crystal lattice. Metals are almost always crystalline and the individual crystal grains are typically micron size. This means that the spatial extent of the orbitals is *very large* compared to the size of the atoms or the unit cell. The diagram at the left shows a generic plot of electron energy vs. density of states for a metal such as Na, Cu, or Ag. In these cases, there are N orbitals for N electrons, and each orbital can accommodate two electrons. Therefore the Fermi level, which corresponds to the energy of the highest occupied MO at zero temperature, is somewhere in the middle of the band of orbitals. The energy level spacing between orbitals is very small compared to the thermal energy kT, so we can think of the orbitals as forming a continuous band.

Classically, if the electrons in this band were free to be thermally excited, we would expect them to have a specific heat of 3R per mole of electrons. However, experimentally we observe that C_p is only about 0.02 R per mole. This suggests that only about 1% of the electrons in the metal can be thermally excited at room temperature. However, essentially all of the valence electrons are free to move in the crystal and contribute to electrical conduction. To understand this apparent paradox, we need to recall that the electrons exist in quantized energy levels.

Because of quantization, electrons in metals have a Fermi-Dirac distribution of energies. In this distribution, most of the electrons are spin-paired, although the individual electrons in these pairs can be quite far apart since the orbitals extend over the entire crystal. A relatively small number of electrons at the top of the Fermi sea are unpaired by thermal excitation. This is the origin of the Pauli paramagnetism of metals.

How fast are electrons traveling in a typical metal? Because of the bell shape of the E vs. DOS curve, most of the electrons have $E \approx E_F$. At the midpoint energy (E_F) of the band, the MO's have one node for every two atoms. We can calculate the de Broglie wavelength as twice the distance between nodes and thus:

$\lambda = 4a$ at the midpoint of the band.

where a is the interatomic spacing. Since a typical value of a is about 2 Å, we obtain the de Broglie wavelength $\lambda \approx 8$ Å.

Using the de Broglie relation $p = h/\lambda$, we can write:

$$p = h/\lambda = m_e v_F,$$

where m_e is the mass of the electron and v_F is the velocity of electrons with energy E_F.

Solving for $v_F = h/m_e\lambda$ we obtain $v_F = (6.62 \times 10^{-34}\ \text{J s})/(9.1 \times 10^{-31}\ \text{kg})(8 \times 10^{-10}\ \text{m}) = 1.0 \times 10^{6}$ m/s, which is about 1/300 the speed of light. This is the Fermi velocity.

Experimental values of v_F are 1.07×10^6 and 1.39×10^6 m/s for Na and Ag, respectively, confirming the above estimate.

However, the drift velocity of electrons in metals - the average speed at which electrons move in applied electric field - is quite slow, on the order of 0.0001 m/s, or .01 cm/s.

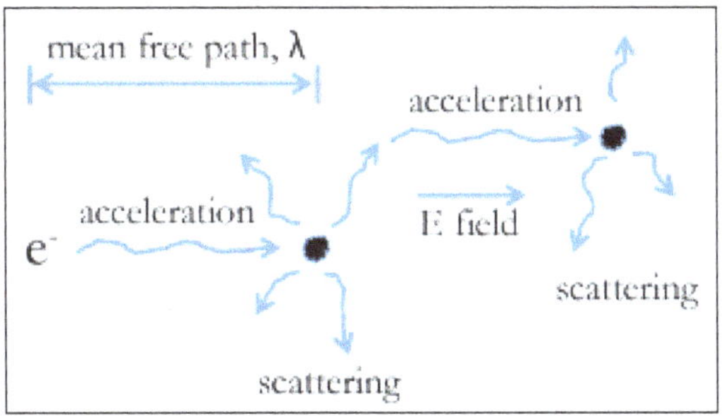

In order to understand the great disparity between the Fermi velocity and the drift velocity of electrons in metals, we need to consider a picture for the scattering of electrons, and their acceleration in an electric field, as shown at the left. If we apply a voltage across a metal (e.g., a metal wire), the electrons are subjected to an electric field E, which is the voltage divided by the length of the wire. This electric field exerts a force on the electron, causing it to accelerate. However, the electron is frequently scattered, mostly by phonons (lattice vibrations). Each time the electron is scattered its acceleration starts all over again. The time between scattering events is τ and the distance the electrons travel between scattering events is the mean free path, λ. (This is NOT the same λ as the de Broglie wavelength, they just are being denoted by the same symbol.)

We can write the force on the electron as:

$$F = eE = m_e a = m_e v_{drift}/\tau$$

For ordinary metals, the mean free path of a valence electron (λ) is quite long relative to the interatomic spacing, represented in this analogy as the length of a football. In this equation, a is the acceleration in the electric field, m_e is the mass, and v_{drift} is the drift velocity of the electron.

Experimentally, the mean free path is typically obtained by measuring the scattering time. For an electron in Cu metal at 300 K, the scattering time τ is about 2×10^{-14} s. From this we can calculate the mean free path as:

$$\lambda = v_{avg}\tau \approx v_F\tau = (1 \times 10^6 \text{ m/s})(2 \times 10^{-14} \text{ s}) = 40 \text{ nm}.$$

The mean free path (40 nm = 400 Å) is quite long compared to the interatomic spacing (2 Å). To put it in perspective, if the interatomic spacing were scaled to the length of a football (0.3 m), the mean free path would be over half the length of the football field (60 m). Thus an electron travels a fairly long way between scattering events and scarcely notices the atomic structure of the metal in which it is traveling.

Electrons are traveling in metals at the Fermi velocity v_F, which is very, very fast (10^6 m/s), but the flux of electrons is the same in all directions. That is, they are *going nowhere fast*. In an electric field, a very small but directional *drift velocity* is superimposed on this fast random motion of valence electrons.

We can calculate the drift velocity of electrons as the acceleration in the electric field times the scattering time:

From F = ma, we obtain the acceleration (a) as:

$$a = F/m_e = eE/m_e$$

and thus,

$$v_{drift} = a\tau = eE\tau/m_e$$

If we divide both sides of this equation by the magnitude of the electric field (E), we obtain the mobility (μ):

$$\mu = v_{drift}/E = e\tau/m_e$$

μ has units of velocity/field = cm/s / V/cm = cm^2/Vs.

An important consequence of the calculation of v_{drift} is Ohm's Law, V = iR. From the equations above, we can see that the drift velocity increases linearly with the applied electric field. The drift velocity (cm/s) is proportional to the current (i, coul/s), and the electric field (E, V/cm) is proportional to the voltage (V):

$$\text{Current (i)} = n\, e\, v_{drift} \times \text{area}$$

$$\text{Voltage (V)} = E \times \text{length}$$

Here n is the density of valence electrons ($\#/cm^3$) and e is the charge of the electron (coul). Combining these equations with our equation for v_{drift} we obtain:

$$V = i\left(\frac{m_e}{ne^2\tau}\right)\left(\frac{\text{length}}{\text{area}}\right) = iR$$

Thus, $V = iR$, where R is the combination of the two terms in parentheses. The first of these is the resistivity, ρ, and the second is a geometrical factor.

The conductivity (σ) of a metal, which is the inverse of ρ, is proportional to μ, which in turn is proportional to τ (and λ):

$$\sigma = ne\mu = ne^2\tau/m_e$$

We can use this equation to work out the conductivity of a specific metal (Cu), for which $n = 8.5 \times 10^{22}$ cm^{-3} and $\tau = 2 \times 10^{-14}$ s. Putting in the numbers for m_e and e, we obtain $\sigma = 7 \times 10^5\ \Omega^{-1}$ cm^{-1} for Cu, in good agreement with the measured value ($6 \times 10^5\ \Omega^{-1}$cm^{-1}).

Atomic Orbitals and Magnetism

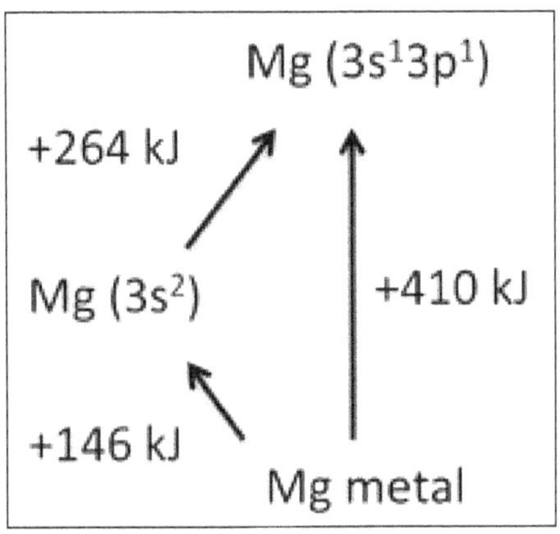

The cohesive energy of Mg metal is the difference between the bonding and promotion energies. The ground state of a gas phase Mg atom is [Ne]$3s^2$, but it can be promoted to the [Ne]$3s^1 3p^1$ excited (gas) state, which is 264 kJ/mol above the ground state. Mg uses two electrons per atom to make bonds, and the sublimation energy of the metal is 146 kJ/mol to move from solid to gas phase.

Mg crystals.

The MO picture helps us rationalize the electrical conductivity of Na ($3s^1$), but what about Mg, which (as an atom in the gas phase) has a $3s^2$ electronic configuration? The two valence electrons are spin-paired in atomic Mg, as they are in the helium atom

(1s²). When the 3s orbitals of Mg combine to form a band, we would expect the band to be completely filled, since Mg has two electrons per orbital. By this reasoning, solid Mg should be an insulator. But Mg has all the properties of a metal: high electrical and thermal conductivity, metallic luster, malleability, etc. In this case the 3s and 3p bands are sufficiently broad (because of strong orbital overlap between Mg atoms) that they form a continuous band. This band, which contains a total of four orbitals (one 3s and three 3p) per atom, is only partially filled by the two valence electrons.

Another way to think about this is to consider the hybridization of the 3s and 3p electrons in Mg. Hybridization requires promotion from the $3s^2 3p^0$ ground state of an Mg atom to a $3s^1 3p^1$ excited state. The promotion energy (+264 kJ/mol) is more than offset by the bonding energy (-410 kJ/mol), the energy released when gaseous atoms in the excited state condense to form the metallic solid. The heat of vaporization, or the cohesive energy of a metal, is the difference between the bonding energy and the promotion energy. Experimentally, we can measure the vaporization energy (+146 kJ/mol) and the promotion energy and use them to calculate the bonding energy. From this we learn that each s or p electron is worth about 200 kJ/mol in bonding energy. The concepts of promotion energy and bonding energy are very useful in rationalizing periodic trends in the bond strengths and magnetic properties of metals.

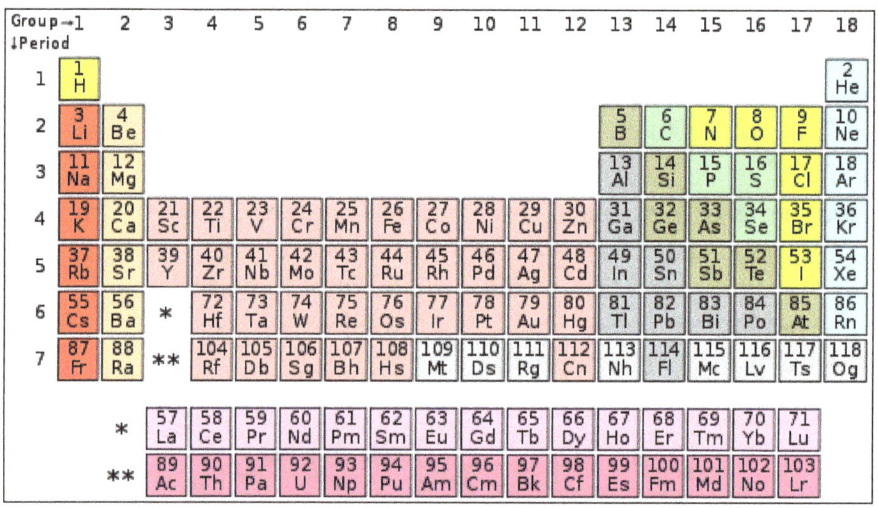

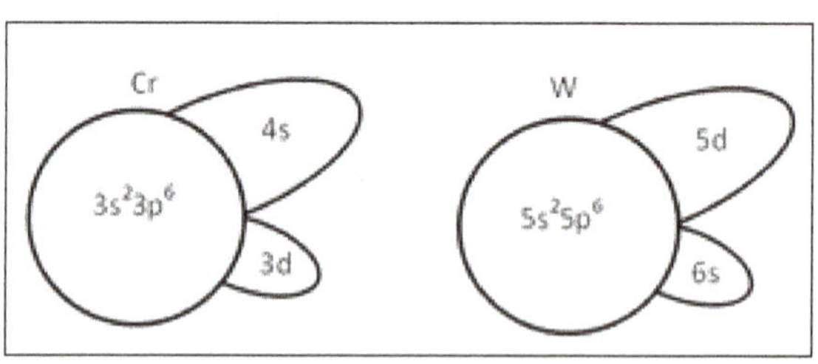

Schematic representation of the sizes of different orbitals of Cr and W.

In the first transition series, shielding of the 3d orbitals is poor. Therefore the 4s and 4p orbitals are more effective in bonding than the 3d. In the third transition series, the situation is reversed. The increased nuclear charge is felt most strongly by the 6s, which takes on the character of an inert electron pair. The 5d orbitals are well shielded by the complete n=4 shell so they have good orbital overlap with neighboring atoms.

Periodic trends in d-electron bonding. While electrons in s and p orbitals tend to form strong bonds, d-electron bonds can be strong or weak. There are two important periodic trends that are related to orbital size and orbital overlap. As we move *across the periodic table* (Sc - Ti - V - Cr - Fe), the *d orbitals contract* because of increasing nuclear charge. Moving *down the periodic table* (V - Nb - Ta), the *d orbitals expand* because of the increase in principal quantum number. These trends explain the distinct behavior of the 3d elements relative to those in the 4d and 5d series. In the 3d series, the contraction of orbitals affects the ability of the d electrons to contribute to bonding. Past V in the first row of the transition metals, the 3d electrons become much less effective in bonding because they overlap weakly with their neighbors. Weak overlap of 3d orbitals gives narrow d-bands and results in the emergence of magnetic properties.

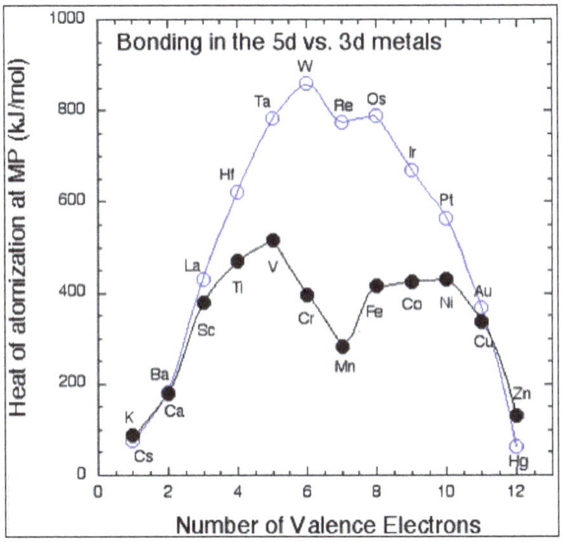

The heat of vaporization (the cohesive energy) of metals in the 3d and 5d series, measured at the melting point of the metal.

In the 4d and 5d series, a plot of cohesive energy vs. number of valence electrons (below left) has a "volcano" shape that is peaked at the elements Mo and W ($5s^1 4d^5$ and $6s^1 5d^5$, respectively). The number of bonding electrons, and therefore the bonding energy, increases steadily going from Rb to Mo in the 4d series, and from Cs to W in the 5d series. Mo and W have the most bonding energy because they can use all six of their valence electrons in bonding without promotion. Elements past Mo and W have more d electrons, but some of them are spin paired and so some promotion energy is needed to prepare these electrons for bonding. For example, Pt metal must be promoted from the $6s^1 5d^9$ atomic ground state to $6s^1 5d^7 6p^2$ in order to make six bonds per atom, and the energy cost of promoting electrons from the 5d to the 6p orbitals is reflected in the net bonding energy. Because of their strong bonding energy, elements in the middle of the 4d and 5d series have very high melting points. We do not see magnetism in the 4d or 5d metals or their alloys because orbital overlap is strong and the bonding energy exceeds the electron pairing energy.

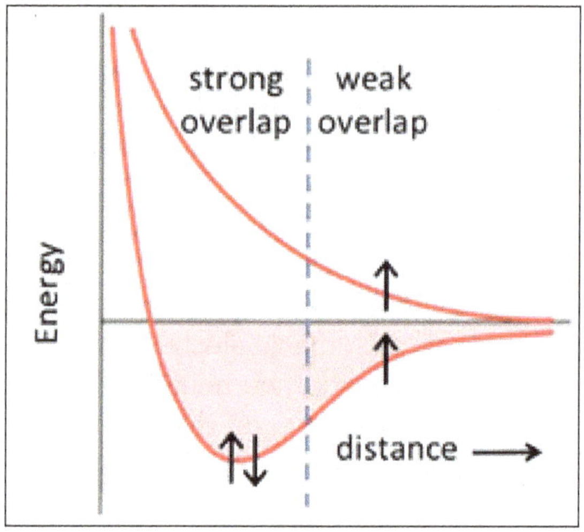

With strong overlap between orbitals of neighboring atoms, the bonding energy exceeds the pairing energy and electrons spin-pair. With weaker overlap, bonding is weak and spins unpair, resulting in magnetic behavior.

The 3d elements (Sc through Zn) are distinctly different from the 4d and 5d elements in their bonding (and consequently in their magnetic properties). In the 3d series, we see the expected increase in cohesive energy going from Ca ($4s^2$) to Sc ($4s^23d^1$) to Ti ($4s^23d^2$) to V ($4s^23d^3$), but then something very odd happens. The 3d series has a "crater" in the cohesive energy plot where there was a peak in the 5d series. The cohesive energy actually *decreases* going from V to Mn, even though the number of valence electrons is increasing. We can explain this effect by remembering that the 3d orbitals are progressively contracting as more protons are added to the nucleus. For elements beyond V, the orbital overlap is so poor that the 3d electrons are no longer effective in bonding, and the valence electrons begin to unpair. At this point the elements become magnetic. Depending on the way the spins order, metals and alloys in this part of the periodic table can be ferromagnetic (spins on neighboring atoms aligned parallel, as in the case of Fe or Ni) or antiferromagnetic (spins on neighboring atoms antiparallel, as in the case of Mn). We have seen the trade-off between orbital overlap and magnetism before in the context of paramagnetic transition metal complexes. It is worth recalling that this behavior is predicted in the energy vs. distance diagram of the hydrogen atom. At short interatomic distances (or with strong overlap between atomic orbitals), the spins of the electrons pair and a bond is formed. Unpairing the electrons becomes favorable at larger interatomic distances where the overlap between orbitals is poor.

Interestingly, many alloys of the 4f elements (the lanthanides) are also magnetic because the 4f orbitals, like the 3d orbitals, are poorly shielded from the nuclear charge and are ineffective in bonding. Strong permanent magnets often contain alloys of Nd, Sm, or Y, usually with magnetic 3d elements such as Fe and Co. Because the 4f orbitals are contracted and not very effective in bonding, other physical properties of the lanthanides can also be affected. For example, it has been proposed that oxides of the 4f elements have weak surface interactions with polar molecules such as water because of f-orbital contraction. Experimentally, CeO_2, Er_2O_3, and Ho_2O_3 are observed to be hydrophobic, whereas main group and early transition metal oxides (e.g., Al_2O_3, SiO_2, TiO_2) are quite hydrophilic.

Late transition metal alloys. Although the bonding in the 5d series follows a "normal" volcano plot, the situation is a bit more complex for alloys of Re, Os, Ir, Pt, and Au. There is strong overlap between the 5d orbitals, but because these elements contain more than five d-electrons per atom, they cannot make as many bonds as 4d or 5d elements with half-filled d-shells such as Mo or W. This progressive filling of the d-band explains the steady decrease in bonding energy going from Os to Au. Pt and Au are both soft metals with relatively low heats of vaporization. However, these metals (especially Ir, Pt, and Au) can combine with early transition metals to form stable alloys with very negative heats of formation and high melting points. For example, ZrC and Pt react to form a number of stable alloys (ZrPt, Zr_9Pt_{11}, Zr_3Pt_4, $ZrPt_3$) plus carbon. This reactivity is unusual because we normally think of Pt as a "noble" (i.e., unreactive) metal, and because ZrC is a very stable, refractory metal carbide. The favorable combination of early and late transition metals has been interpreted as arising from a d-electron "acid-base" interaction. For example, in $HfPt_3$, Hf is the "acid" with an electron configuration of $6s^25d^2$, while Pt is the "base" with the electron configuration of $6s^15d^9$. They combine to create a stable "salt" product with a filled 5d electron configuration without promoting any electrons to higher orbitals. The implication is that Pt donates d-electrons to the "d-acid," Zr or Hf. However, electronic structure calculations on model compounds show that much of the bonding energy in ZrPt and $ZrPt_3$ arises from electron transfer from Zr to Pt (not the other way around) and the polarity of the resulting metal-metal bonds. Regardless of the source of their stability, some early-late transition metal alloys are of particular interest for use in catalysis. For example, the alloy $ScPt_3$ is a good catalyst for oxygen reduction in fuel cells. Even though Sc is an active metal that is easily oxidized, it is stabilized in the aqueous acid environment of the fuel cell by its strong interaction with Pt.

Filling of the 3d and 4s,4p bands: In the 3d series, we see magnetic behavior for elements and alloys between Cr and Ni. Past Ni, the elements (Cu, Zn, Ga,) are no longer magnetic and they are very good electrical conductors, implying that their valence electrons are highly delocalized. We can understand this behavior by considering the overlap of 4s, 4p, and 3d orbitals, all of which are close in energy. The 4s and 4p have strong overlap and form a broad, continuous band. On the other hand, the 3d electrons are contracted and form a relatively narrow band. Progressing from the early 3d elements (Sc, Ti, V), we begin to fill the 3d orbitals, which are not yet so contracted that they cannot contribute to bonding. Thus, the valence electrons in Sc, Ti, and V are all spin-paired, except for a small number near the Fermi level that give rise to a weak Pauli paramagnetism. Moving across the 3d series to the magnetic elements (Fe, Co, Ni), the d-orbitals are now so contracted that their electrons unpair and we see cooperative ordering of spins (ferromagnetism and antiferromagnetism). Referring to the band diagram at the right, the 3d band is only partially filled and the Fermi level cuts through it. For Cu, Zn, and Ga, the 3d orbitals are even more contracted and the 3d band is thus more narrow, but now it is completely filled and the Fermi level is in the 4s,4p band. The strong orbital overlap in these bands results in spin pairing and a high degree of electron delocalization. Consequently, metals in this part of the periodic table (Cu, Ag) are diamagnetic and are among the best electrical conductors at room temperature. Finally, at Ge, the 4s,4p band is completely filled and the solid is a semiconductor.

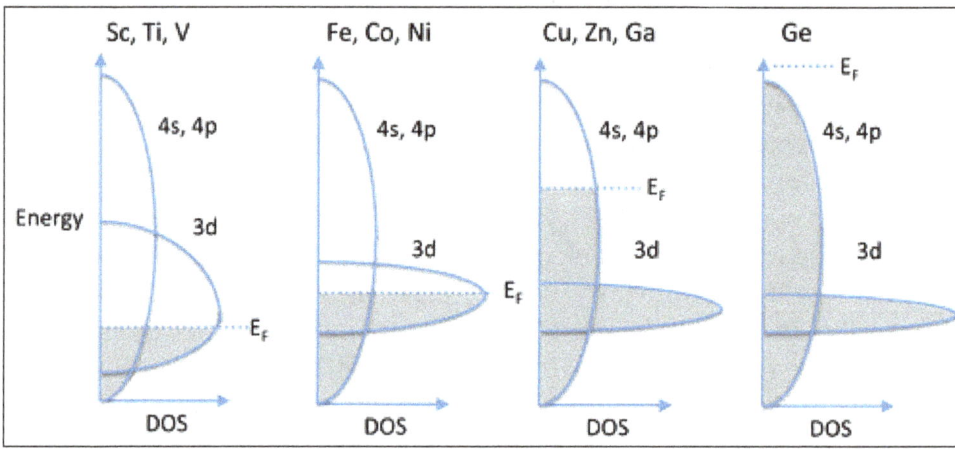

Progressive filling of the 3d and 4s,4p bands going across the periodic table from Sc to Ge.

Materials are classified as diamagnetic if they contain no unpaired electrons. Diamagnetic substances are very weakly repelled from an inhomogeneous magnetic field, molecules or ions that have unpaired spins are paramagnetic and are attracted to a magnet, i.e., they move towards the high field region of an inhomogeneous field. This attractive force results from the alignment of spins with the field, but in the case of paramagnetism each molecule acts independently. In metals, alloys, oxides, and other solid state compounds, the unpaired spins interact strongly with each other and can order spontaneously, resulting in the cooperative magnetic phenomena.

Ferro-, Ferri- and Antiferromagnetism

The magnetism of metals and other materials are determined by the orbital and spin motions of the unpaired electrons and the way in which unpaired electrons align with each other. All magnetic substances are paramagnetic at sufficiently high temperature, where the thermal energy (kT) exceeds the interaction energy between spins on neighboring atoms. Below a certain critical temperature, spins can adopt different kinds of ordered arrangements.

Ferromagnetic ↑ ↑ ↑ ↑ ↑ ↑ ↑	Below T_C, spins are aligned parallel in magnetic domains
Antiferromagnetic ↑ ↓ ↑ ↓ ↑ ↓ ↑	Below T_N, spins are aligned antiparallel in magnetic domains
Ferrimagnetic ↑ ↓ ↑ ↓ ↑ ↓ ↑	Below T_C, spins are aligned antiparallel but do not cancel
Paramagnetic ↑ ↓ ↑ ↑ ↓ ↑ ↓	Spins are randomly oriented (any of the others above T_C or T_N)

A pictorial description of the ordering of spins in ferromagnetism, antiferromagnetism, ferrimagnetism, and paramagnetism.

Let's begin by considering an individual atom in the bcc structure of iron metal. Fe is in group VIIIb of the periodic table, so it has eight valence electrons. The atom is promoted to the $4s^13d^7$ state in order to make bonds. A localized picture of the d-electrons for an individual iron atom might look like this:

↑↓ ↑↓ ↑ ↑ ↑

Since each unpaired electron has a spin moment of 1/2, the total spin angular momentum, S, for this atom is:

$$S = 3(1/2) = 3/2 \text{ (in units of } h/2\pi)$$

We can think of each Fe atom in the solid as a little bar magnet with a spin-only moment S of 3/2. The spin moments of neigboring atoms can align in parallel (↑ ↑), antiparallel (↑ ↓), or random fashion. In bcc Fe, the tendency is to align parallel because of the positive sign of the exchange interaction. This results in ferromagnetic ordering, in which all the spins within a magnetic domain (typically hundreds of unit cells in width) have the same orientation, as shown in the figure at the right. Conversely, a negative exchange interaction between neighboring atoms in bcc Cr results in antiferromagnetic ordering. A third arrangement, ferrimagnetic ordering, results from an antiparallel alignment of spins on neighboring atoms when the magnetic moments of the neighbors are unequal. In this case, the spin moments do not cancel and there is a net magnetization. The ordering mechanism is like that of an antiferromagnetic solid, but the magnetic properties resemble those of a ferromagnet. Ferrimagnetic ordering is most common in metal oxides.

- Magnetization and susceptibility: The magnetic susceptibility, χ, of a solid depends on the ordering of spins. Paramagnetic, ferromagnetic, antiferromagnetic, and ferrimagnetic solids all have $\chi > 0$, but the magnitude of their susceptibility varies with the kind of ordering and with temperature. We will see these kinds of magnetic ordering primarily among the 3d and 4f elements and their alloys and compounds. For example, Fe, Co, Ni, $Nd_2Fe_{14}B$, $SmCo_5$, and YCo_5 are all ferromagnets, Cr and MnO are antiferromagnets, and Fe_3O_4 and $CoFe_2O_4$ are ferrimagnets. Diamagnetic compounds have a weak *negative* susceptibility ($\chi < 0$).
    - H = applied magnetic field (units: Henry (H)).
    - B = induced magnetic field in a material (units: Tesla (T)).
    - M = magnetization, which represents the magnetic moments within a material in the presence of an external field H.
    - Magnetic susceptibility $\chi = M/H$.

Usually, χ is given in *molar* units in the *cgs system*:

χ_M = molar susceptibility (units: cm^3/mol)

Typical values of χ_M:

Compound	Type of Magnetism	χ at 300K (cm^3/mol)
SiO_2	Diamagnetic	-3×10^{-4}
Pt metal	Pauli paramagnetic	$+2 \times 10^{-4}$
$Gd_2(SO_4)_3 \cdot 8H_2O$	Paramagnetic	$+5 \times 10^{-2}$
Ni-Fe alloy	Ferromagnetic	$+10^4 - 10^6$

To correlate χ with the number of unpaired electrons in a compound, we first correct for the small diamagnetic contribution of the core electrons:

$$\chi^{corr} = \chi^{obs} - \chi^{diamagnetic\ cores}$$

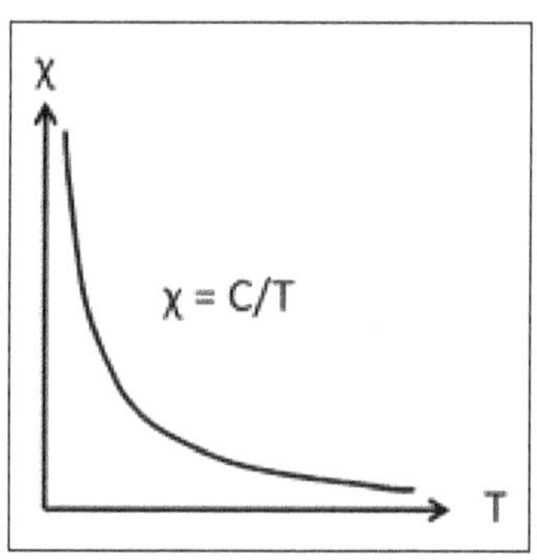

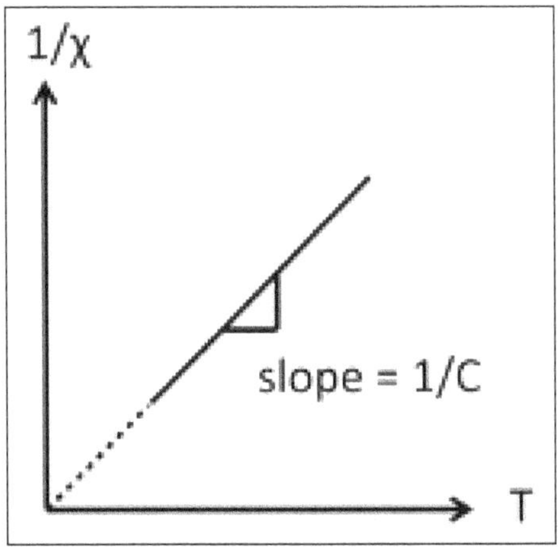

Curie law behavior of a paramagnet. A plot of 1/χ vs. absolute temperature is a straight line, with a slope of 1/C and an intercept of zero.

Susceptibility of paramagnets. For a paramagnetic substance,

$$\chi_M^{corr} = \frac{C}{T}$$

The inverse relationship between the magnetic susceptibility and T, the absolute temperature, is called Curie's Law, and the proportionality constant C is the Curie constant:

$$C = \frac{N_A}{3k_B}\mu_{eff}^2$$

Note that C is not a "constant" in the usual sense, because it depends on μ_{eff}, the effective magnetic moment of the molecule or ion, which in turn depends on its number of unpaired electrons:

$$\mu_{eff} = \sqrt{n(n+2)}\mu_B$$

Here μ_B is the Bohr magneton, a physical constant defined as $\mu_B = eh/4\pi m_e = 9.274 \times 10^{-21}$ erg/gauss (in cgs units).

In cgs units, we can combine physical constants,

$$\frac{N_A}{3k_B}\mu_B^2 = .125$$

Combining these equations, we obtain,

$$\chi_M^{corr} = \frac{.125}{T}\left(\frac{\mu_{eff}}{\mu_B}\right)^2$$

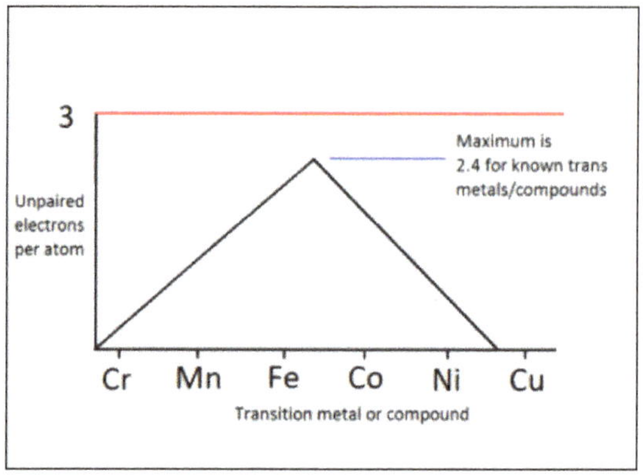

Number of unpaired electrons per atom, determined from Curie constants of transition metals and their 1:1 alloys.

These equations relate the molar susceptibility, a bulk quantity that can be measured with a magnetometer, to μ_{eff}, a quantity that can be calculated from the number of unpaired electrons, n. Two important points to note about this formula are:

- The magnetic susceptibility is inversely proportional to the absolute temperature, with a proportionality constant C (Curie's Law).
- So far we are talking only about paramagnetic substances, where there is no interaction between neighboring atoms.

Returning to the isolated Fe atom with its three unpaired electrons, we can measure the Curie constant for iron metal (above the temperature of its transition to a paramagnetic solid) and compare it to the calculation of μ_{eff}. Since n = 3, we calculate:

$$\mu_{eff} = \sqrt{(3)(5)}\mu_B = 3.87\mu_B$$

The plot at the right shows the number of unpaired electrons per atom, calculated from measured Curie constants, for the magnetic elements and 1:1 alloys in the 3d series. The plot peaks at a value of 2.4 spins per atom, slightly lower than we calculated for an isolated iron atom. This reflects that fact that there is some pairing of d-electrons, i.e., that they do contribute somewhat to bonding in this part of the periodic table.

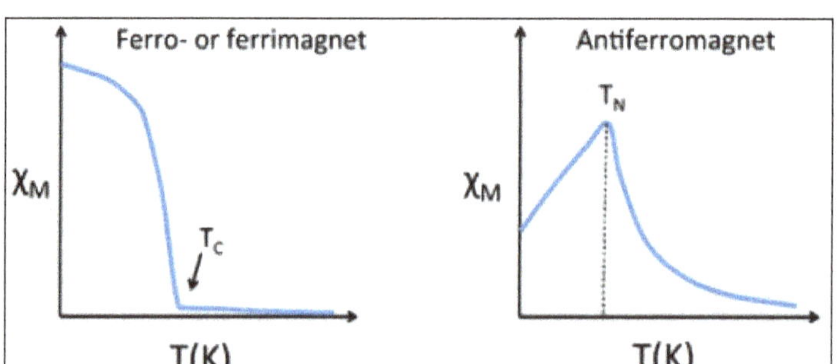

Magnetic susceptibility vs temperature (Kelvin) for ferrimagnetic, ferromagnetic, and antiferromagnetic materials.

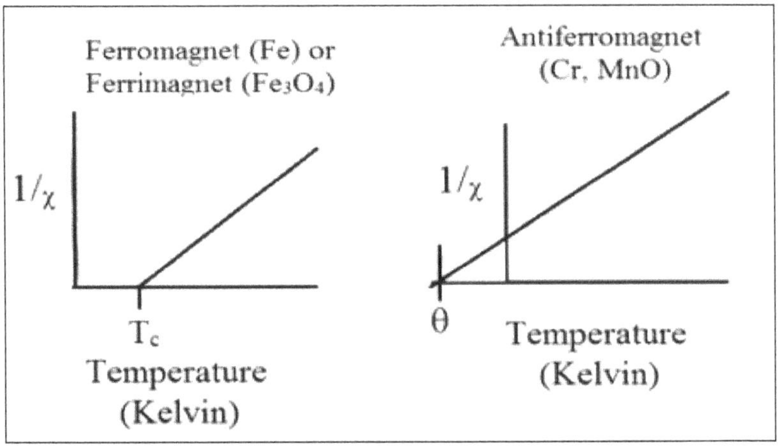

Plots of $1/\chi$ vs. T for ferromagnets, ferrimagnets, and antiferromagnets.

Susceptibility of Ferro-, Ferri-, and Antiferromagnets

Below a certain critical temperature, the spins of a solid paramagnetic substance order and the susceptibility deviates from simple Curie-law behavior. Because the ordering depends on the short-range exchange interaction, this critical temperature varies widely. Metals and alloys in the 3d series tend to have high critical temperatures because the atoms are directly bonded to each other and the interaction is strong. For example, Fe and Co have critical temperatures (also called the Curie temperature, T_c, for ferromagnetic substances) of 1043 and 1400 K, respectively. The Curie temperature is determined by the strength of the magnetic exchange interaction and by the number of unpaired electrons per atom. The number of unpaired electrons peaks between Fe and Co as the d-band is filled, and the exchange interaction is stronger for Co than for Fe. In contrast to ferromagnetic metals and alloys, paramagnetic salts of transition metal ions typically have critical temperatures below 1K because the magnetic ions are not directly bonded to each other and thus their spins are very weakly coupled in the solid state. For example, in gadolinium sulfate, the paramagnetic Gd^{3+} ions are isolated from each other by SO_4^{2-} ions.

Above the critical temperature T_C, ferromagnetic compounds become paramagnetic and obey the Curie-Weiss law:

$$\chi = \frac{C}{T - T_c}$$

This is similar to the Curie law, except that the plot of $1/\chi$ vs. T is shifted to a positive intercept T_C on the temperature axis. This reflects the fact that ferromagnetic materials (in their paramagnetic state) have a greater tendency for their spins to align in a magnetic field than an ordinary paramagnet in which the spins do not interact with each other. Ferrimagnets follow the same kind of ordering behavior. Typical plots of χ vs. T and $1/\chi$ vs. T for ferro-/ferrimagnets are shown at the right.

Antiferromagnetic solids are also paramagnetic above a critical temperature, which is called the Néel temperature, T_N. For antiferromagnets, χ reaches a maximum at T_N and is smaller at higher temperature (where the paramagnetic spins are further disordered by thermal energy) and at lower temperature (where the spins pair up). Typically, antiferromagnets retain some positive susceptibility even at very low temperature because of canting of their paired spins. However the maximum value of χ is much lower for an antiferromagnet than it is for a ferro- or ferrimagnet. The Curie-Weiss law is also modified for an antiferromagnet, reflecting the tendency of spins (in the paramagnetic state above T_N) to resist parallel ordering. A plot of $1/\chi$ vs. T intercepts the temperature axis at a negative temperature, $-\theta$, and the Curie-Weiss law becomes:

Microcrystalline grains within a piece of $Nd_2Fe_{14}B$ (the alloy used in neodymium magnets) with magnetic domains made visible with a Kerr microscope. The domains are the light and dark stripes visible within each grain.

The movement of domain walls in a grain of silicon steel is driven in this movie by increasing the external magnetic field in the "downward" direction, and is imaged using a

Kerr microscope. White areas are domains with their magnetization directed up, dark areas - which eventually comprise the entire grain - are domains with their magnetization directed down.

Ordering of Spins below T_C

Below T_C, the spins align spontaneously in ferro- and ferrimagnets. Complex magnetization behavior is observed that depends on the history of the sample. For example, if a ferromagnetic material is cooled in the absence of an applied magnetic field, it forms a mosaic structure of that each have internally aligned spins. However, neighboring domains tend to align the opposite way in order to minimize the total energy of the system. This is illustrated in the figure at the left for a Nd-Fe-B magnet. The sample consists of 5-10 μm wide crystal grains that can be easily distinguished by the sharp boundaries in the image. Within each grain are a series of lighter and darker stripes (imaged by using the optical Kerr effect) that are ferromagnetic domains with opposite orientations. Averaged over the whole sample, these domains have random orientation so the net magnetization is zero.

When a sample like this one is magnetized (i.e., exposed to a strong magnetic field), the domain walls move and the favorably aligned domains grow at the expense of those with the opposite orientation. This transformation can be seen in real time in the Kerr microscope. The domain walls are typically hundreds of atoms wide, so movement of a domain wall involves a cooperative tilting of spin orientation (analogous to "the wave" in a sports stadium) and is a relatively low energy process.

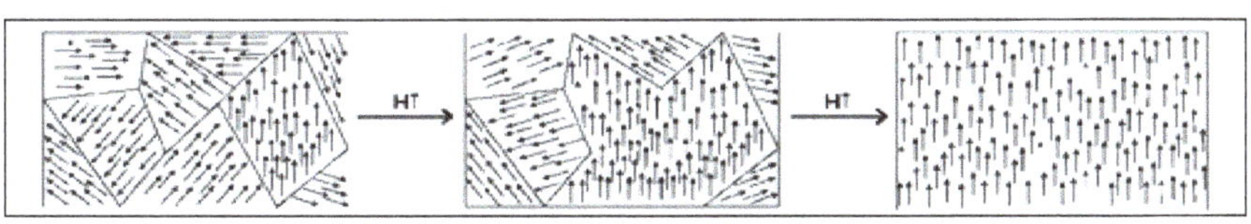

Rotation of orientation and increase in size of magnetic domains in response to an externally applied magnetic field.

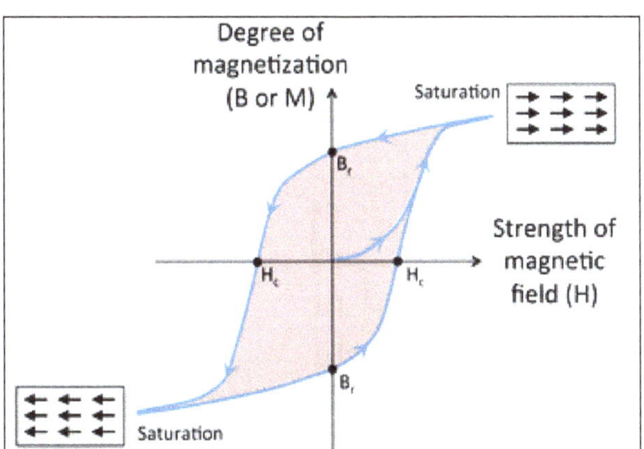

Magnetization of a ferro- or ferrimagnet vs. applied magnetic field H. Starting at the origin, the upward curve is the initial magnetization curve. The downward curve after saturation, along with the lower return curve, form the main loop. The intercepts Hc and Br are the coercivity and remanent magnetization.

The process of magnetization moves the solid away from its lowest energy state (random domain orientation), so magnetization involves input of energy. When the external magnetic field is removed, the domain walls relax somewhat, but the solid (especially in the case of a "hard" magnet) can retain much of its magnetization. If you have ever magnetized a nail or a paper clip by using a permanent magnet, what you were doing was moving the walls of the magnetic domains inside the ferromagnet. The object thereafter retains the "memory" of its magnetization. However, annealing a permanent magnet destroys the magnetization by returning the system to its lowest energy state in which all the magnetic domains cancel each other.

Magnetic Hysteresis

Cycling a ferro- or ferrimagnetic material in a magnetic field results in hysteresis in the magnetization of the material, as shown in the figure at the left. At the beginning, the magnetization is zero, but it begins to rise rapidly as the magnetic field is applied. At high field, the magnetic domains are aligned and the magnetization is said to be saturated. When the field is removed, a certain remanent magnetization (indicated as the point B_r on the graph) is retained, i.e., the material is magnetized. Applying a field in the opposite direction begins to orient the magnetic domains in the other direction, and at a field H_c (the coercive field), the magnetization of the sample is reduced to zero. Eventually the material reaches saturation in the opposite direction, and when the field is removed again, it has remanent magnetization Br, but in the opposite direction. As the field continues to reverse, the magnet follows the hysteresis loop as indicated by the arrows. The area of colored region inside the loop is proportional to the magnetic work done in each cycle. When the field cycles rapidly (for example, in the core of a transformer, or in read-write cycles of a magnetic disk) this work is turned into heat.

Hard and Soft Magnets

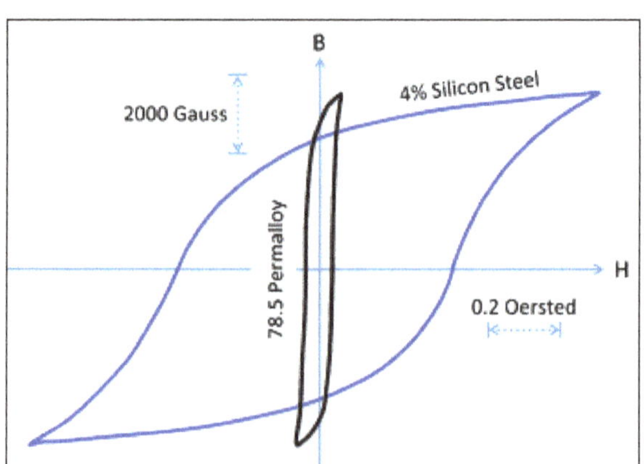

Hysteresis loops comparing a hard magnet (iron-silicon steel) to a soft magnet (permalloy) on the same scale. H_c for permallloy is 0.05 Oe, about 10 times lower than that of the hard magnet. The remanent magnetizations of the two materials are comparable.

Whether a ferro- or ferrimagnetic material is a hard or a soft magnet depends on the strength of the magnetic field needed to align the magnetic domains. This property is characterized by H_c, the coercivity. Hard magnets have a high coercivity (H_c), and thus retain their magnetization in the absence of an applied field, whereas soft magnets have low values.

The figure compares hysteresis loops for hard and soft magnets. Recall that the energy dissipated in magnetizing and demagnetizing the material is proportional to the area of the hysteresis loop. We can see that soft magnets, while they can achieve a high value of B_{sat}, dissipate relatively little energy in the loop. This makes soft magnets preferable for use in transformer cores, where the field is switched rapidly. Permalloy, an alloy consisting of about 20% Fe and 80% Ni, is a soft magnet that has very high magnetic permeability μ (i.e., a large maximum slope of the B vs. H curve) and a very narrow hysteresis loop.

Some materials, such as iron metal, can exist as either hard or soft magnets. Whether bcc iron is a hard or soft magnet depends on the crystal grain size. When crystal grains in iron are sub-micron size, and comparable to the size of the magnetic domains, then the magnetic domains are pinned by crystal grain boundaries. When the magnetic domains are pinned a stronger coercive magnetic field needs to be applied to cause them to re-align. When iron is annealed, the crystal grains grow and the magnetic domains become more free to align with applied magnetic fields. This decreases the coercive field and the material becomes a soft magnet.

Hard magnets such as CrO_2, γ-Fe_2O_3, and cobalt ferrite ($CoFe_2O_4$) are used in magnetic recording media, where the high coercivity allows them to retain the magnetization state (read as a logical 0 or 1) of a magnetic bit over long periods of time. Hard magnets are also used in disk drives, refrigerator magnets, electric motors and other applications. Drive motors for hybrid and electric vehicles such as the Toyota Prius use the hard magnet $Nd_2Fe_{14}B$ (also used to make strong refrigerator magnets) and require 1 kilogram (2.2 pounds) of neodymium. A high-resolution transmission electron microscope image of $Nd_2Fe_{14}B$ is shown below and compared to the crystal structure with the unit cell marked.

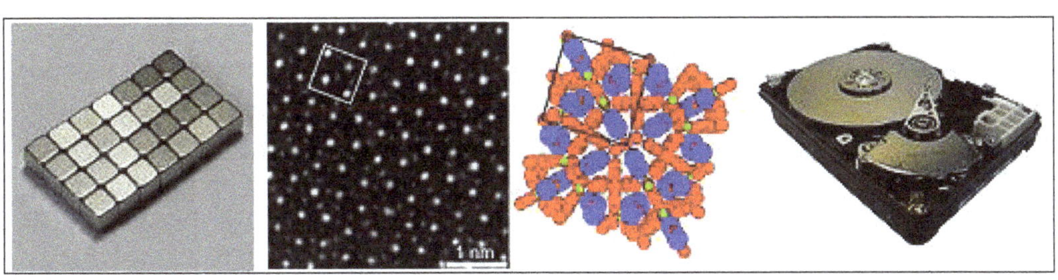

Mechanical Properties of Metals and Alloys

Defects in Metallic Crystals

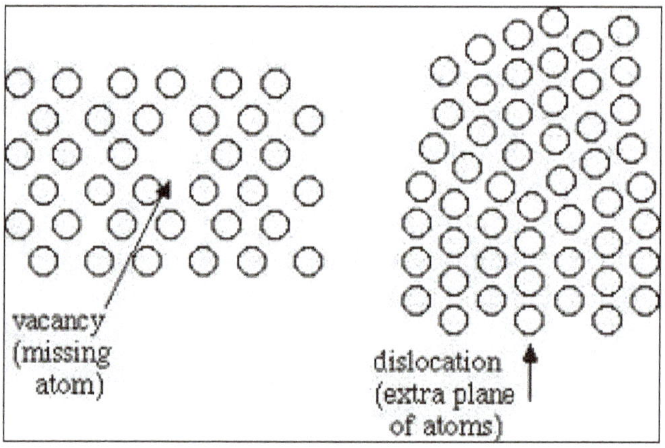

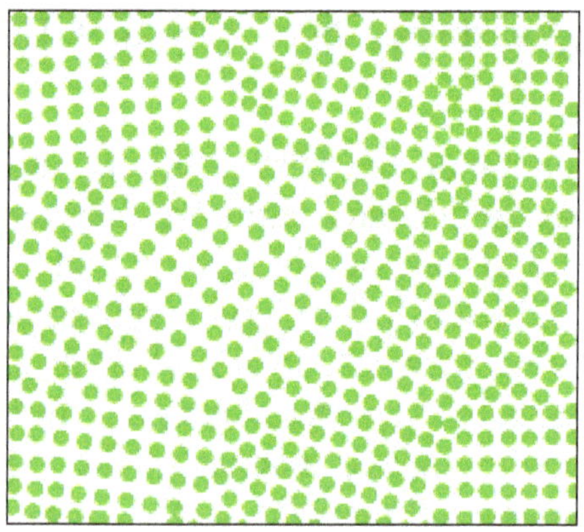

Grains and grain boundaries in a polycrystalline material.

Metals, by virtue of their non-directional bonding, are more energetically tolerant of defects than are covalent network or ionic solids. Because there is no strong preference for one atomic position over another, the energy of a metallic crystal is not greatly impaired by the vacancy of a single atom or by the dislocation of a group of atoms. These kinds of "mistakes" in the packing of metal atoms within crystals are collectively called defects. The deformability of metals is the direct result of defects in the crystal structure. Defects in metals such as Al and Fe are responsible for the three orders of magnitude difference between the yield stress of annealed polycrystalline samples (i.e., normal articles of commerce) and perfect single crystals.

There are several different kinds of defects that can be found in metallic crystals. One kind is called a vacancy, i.e., a place where an atom is missing in the structure. A dislocation, on the other hand, is a line defect; it runs somewhat like a string through the crystal. A dislocation is the result of one atom or group of atoms being pulled slightly out of position with respect to perfect crystal packing. A third kind of defect is called a grain boundary; it is a two-dimensional interface between two different crystal grains in a solid sample. Since the two crystallites have in general different orientation, the structures do not match up exactly at the interface. While point vacancy defects do not significantly affect the mechanical properties of metallic crystals, both dislocations and grain boundaries have large effects.

Work Hardening, Alloying and Annealing

One of the questions we would like to ask is, why are the yield stresses of normal (polycrystalline) metal samples so much lower (by a factor of 1000) than they are in perfect single crystals? The answer has to do with the motion of dislocations. Consider the picture below, which shows planes of metal atoms near a dislocation (the individual atoms are numbered to help you see which bonds are broken and which are formed). The arrows indicate force applied under shear stress. Notice how the dislocation moves by breaking/making metal-metal bonds.

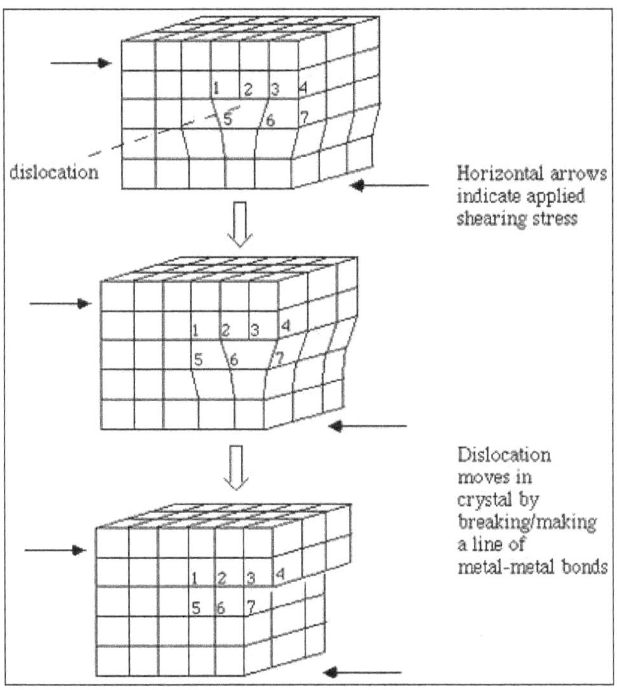

The key point here is that we can induce plastic deformation (shear) by breaking only one line of metal-metal bonds at a time along the dislocation line. This involves far less force than breaking an entire plane of bonds, as we would need to do to shear a perfect crystal. In a given polycrystalline sample, there are many dislocation lines that run perpendicular to all possible shear directions, so their motion can be used to "tear" the metal apart. Turbine rotors on large jets are made of very expensive single crystal nickel-titanium alloys, so that these shearing deformations can be avoided.

We can see that motion of dislocations is basically bad news if we want a metal to be strong and hard (e.g., if we want a structural material, or a knife that can hold a decent edge). There are several ways we can overcome (to some extent) this problem:

- Use single crystals and anneal out all the dislocations (expensive - especially with large items like turbine blades, and impossible with very large items like airplane wings or bridges).
- Work hardening of the metal - this moves all dislocations to grain boundaries (the dislocation essentially becomes part of the grain boundary). Since a grain boundary is a planar defect, it is much less responsive to stress than a line defect.
- Introduce impurity atoms (that is alloying elements) or impurity phases that "pin" the motion of defects. An impurity atom stops the motion because it is a different size, or makes stronger bonds, than the other metal atoms; the line defect has a hard time moving away from rows of such atoms. An impurity phase (like Fe_3C in iron) makes extra grain boundaries that can stop the motion of defects. This effect is analogous to the graphite fibers in fiber-reinforced cross-linked polymers (used, e.g., in tennis rackets) that stop the propagation of cracks.

A simple illustration of work hardening can be done with a piece of copper wire. When struck many times with a hammer, the copper wire becomes stiffer, and it is possible to hang a weight from it. Dislocations move to the crystal grain boundaries during work hardening, effectively halting their motion and at the same time making the individual crystal grains smaller. Because the crystal grains are now smaller, the amount of grain boundary area has increased, and with it the free energy of the material. Annealing reverses the process by lowering the free energy. When the wire is annealed in a flame (heated so that atoms can move and rearrange), the crystal grains grow, and the dislocations reappear. The copper again becomes ductile, and bends easily. Cold-working (work hardening) of metals is important for strengthening structural materials (e.g., iron beams) and for making brittle, hard edges (this is why blacksmiths hammer on knives and swords when they are making them. If you have ever watched them, they do the same thing to horseshoes, when they cool down, to make them stiff).

Malleability of Metals and Alloys

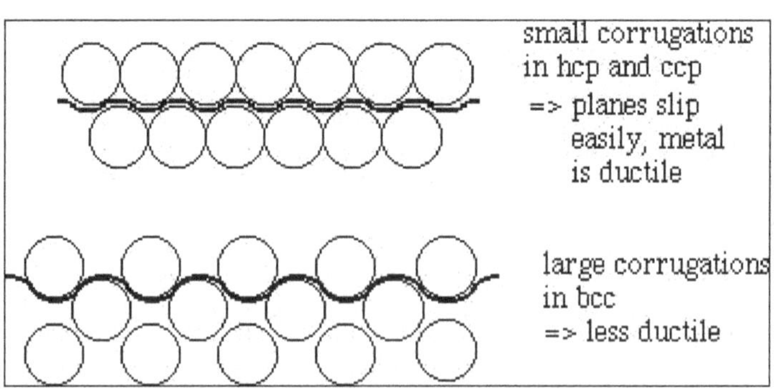

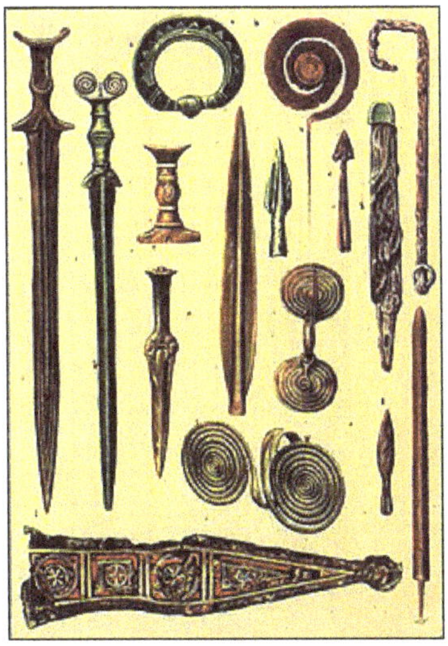

Bronze age weapons from Romania.

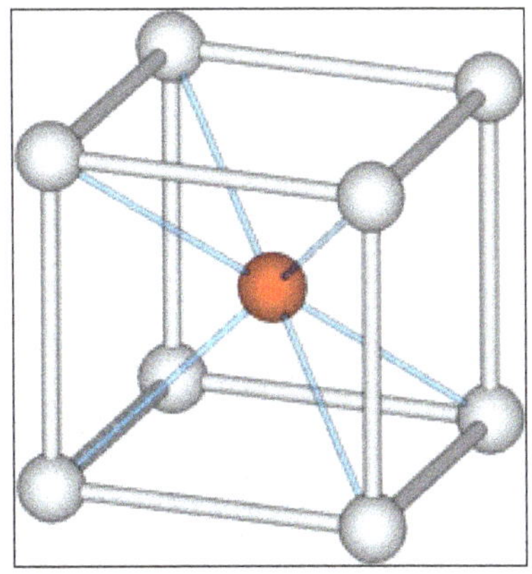

β-brass is an ordered BCC alloy of Cu and Zn.

Metals with close-packed structures (HCP and FCC) such as copper, gold, silver, zinc, magnesium, etc. are in general more malleable than those with the BCC structure (tungsten, vanadium, chromium, etc.). Why? In the close-packed structure, there is relatively little corrugation between sheets of metal atoms. This means that these planes can slip past each other relatively easily. In the BCC structure, there are no close-packed planes, and much greater corrugation between atoms at different levels. This makes it much harder for one row to slide past another.

This effect explains the hardness of alloys like brass (CuZn, which has the BCC structure), which are made by combining two soft metals (Cu and Zn, which are respectively FCC and HCP as pure metals, are both soft and ductile). Bronzes - originally made as alloys of copper and arsenic, but later as alloys of copper and tin - are harder than either of the constituent metals for the same reason.

The history of bronzes and brass dates to pre-historic times, with the earliest brasses made by smelting copper-zinc ores. In the Bronze Age, possession of these hard alloys provided a tactical advantage in warfare, that was later supplanted when the technology for smelting iron was developed.

Iron and Steel

The iron–iron carbide ($Fe–Fe_3C$) phase diagram. Below 912 °C, pure iron exists as the alpha phase, ferrite, which has the BCC structure. Between 912 and 1,394 °C, pure iron exists as the gamma phase, austenite, which has the FCC structure. Carbon is more soluble in the FCC phase, which occupies area "γ" on the phase diagram, than it is in the BCC phase. The percent carbon determines the type of iron alloy that is formed upon cooling from the FCC phase, or from liquid iron: alpha iron, carbon steel (pearlite), or cast iron.

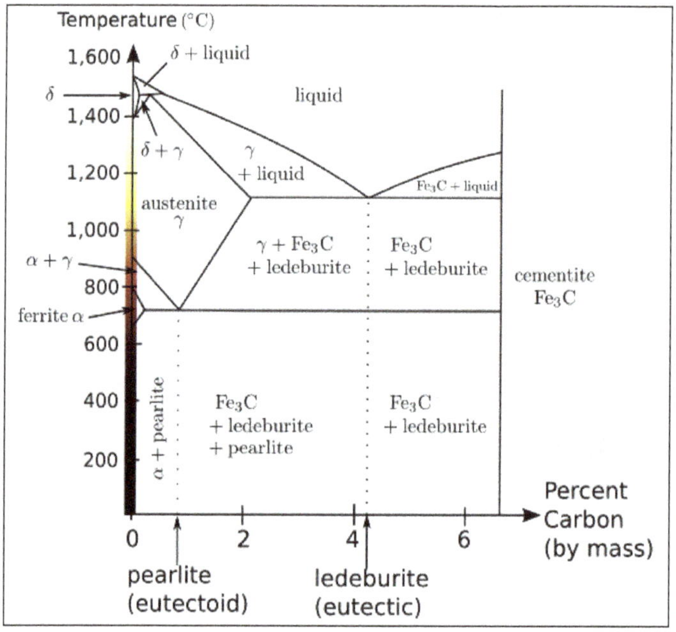

One other very important place where the difference between the hardness of a BCC and a close-packed metal is important is in steelmaking. Between room temperature and 912°C, iron has the BCC structure, and is a tough, hard metal ("tough as nails"). Above 912 °C, pure iron switches over to the FCC (austenite) structure, which is much more ductile. So hot iron can be bent and worked into a variety of shapes when it is very hot but still solid (it melts at 1535 °C). Rapid quenching of hot iron - e.g., when the blacksmith plunges a red hot piece directly into cold water - cools it to room temperature, but doesn't allow time for the FCC --> BCC phase transition to occur; therefore, such pieces are still relatively malleable and can be shaped.

Upon cooling, high carbon steels phase segregate into a mixture of bcc iron (light gray) and Fe3C (dark gray) microscopic grains.

Carbon is added (about 1% by weight) to iron to make "carbon steel", which is a very hard material. Carbon is rather soluble in the FCC phase of iron, but not in the BCC phase. Therefore, when the ductile FCC phase cools and turns into BCC ("tempering" the steel, which means cooling it slowly enough so the FCC to BCC transformation can occur), the iron can no longer dissolve the excess carbon. The carbon forms layers or grains of an extra phase, Fe_3C ("cementite" - a very hard material) which are layered or dotted throughout the matrix of BCC iron grains. The effect of all these little grains of Fe_3C is to stop the motion of dislocations, making for a harder but (with higher carbon content) increasingly brittle material. This is why knives and swords are quenched from the FCC phase, cold worked into the appropriate shapes, and then heated up again and tempered (before they are sharpened) when they are made. Cast iron objects (frying pans, radiators, etc) have a higher carbon content and are therefore very strong, but tend to fracture rather than bend because of the larger fraction of the brittle Fe_3C phase in the alloy.

Amorphous Alloys

Amorphous metals, also called bulk metallic glasses, are materials of growing technological importance. Because their glassy structures do not support the movement of dislocations, they are stronger and more wear-resistant than crystalline metals of similar composition.

All pure metals and most simple alloys crystallize easily. Bulk samples of metals prepared by ordinary chemical or electrochemical methods are polycrystalline, with grain sizes ranging from tens of nanometers to tens of microns in size. Larger crystals of metals can be made by very slow crystal growth, e.g., by the Czochralski process. In contrast, glassy or amorphous metals can be prepared by very rapid cooling from the melt. With pure metals and simple alloys the cooling rate needed is so fast - on the order of 10^6 K/s - that amorphous samples can only be made as very thin films.

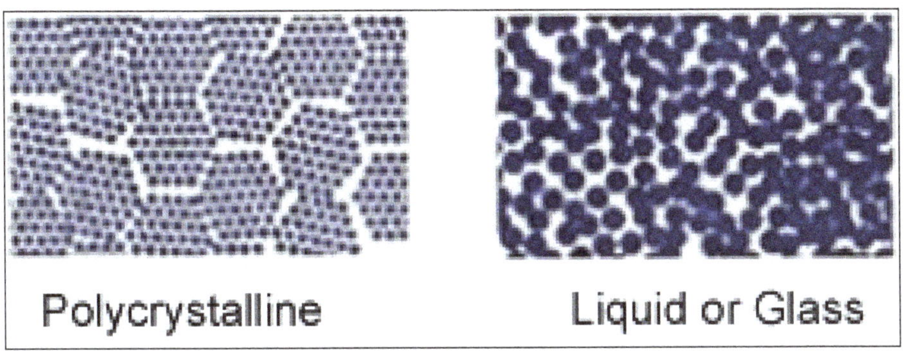

Alloys of metals with more complex stoichiometries can be made in amorphous form by slower cooling from the melt. These alloys have been prepared and studied since the 1960s, and since the 1990s amorphous alloys have been discovered that can be prepared in bulk form at cooling rates on the order of 1 deg/s, similar to the cooling rates of other kinds of glasses.

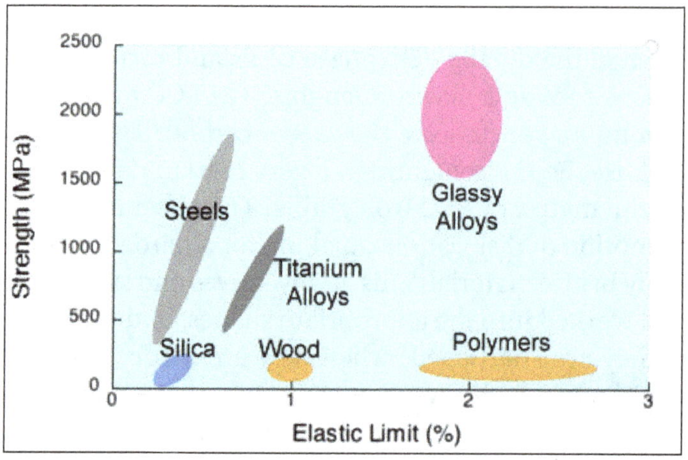

Currently amorphous metals (marketed under the tradenames Vitreloy and Liquid-metal) are used commercially in golf clubs, watches, USB flash drives, and other applications where very high elasticity, yield strength, and/or wear resistance are needed.

Year	Alloy	Cooling Rate (K/s)
1960	$Au_{75}Si_{24}$	10^6 - thin films & ribbons
1969	Pd-Cu-Si	100-1000
1980s	La-Al-Cu & others	1-100
1990s	Zr-Ti-Cu-Ni-Be	~1 (similar to oxide glasses)

References

- Inorganic-chemistry: byjus.com, Retrieved 20, January 2020
- Inorganic-compound, science: britannica.com, Retrieved 12, August 2020
- Electron-Transfer-Reaction: mgcub.ac.in, Retrieved 04, May 2020

5

Physical Chemistry

The field of study which deals with macroscopic and particulate phenomena in chemical systems is called physical study. It focuses on their principles and practices as well as the concepts of physics like motion, force, energy, statistical mechanics, etc. All the diverse aspects of physical chemistry such as chemical kinetics, equilibrium chemistry and thermodynamics have been carefully analyzed in this chapter.

Physical chemistry is a Branch of chemistry concerned with interactions and transformations of materials. Unlike other branches, it deals with the principles of physics underlying all chemical interactions (e.g., gas laws), seeking to measure, correlate, and explain the quantitative aspects of reactions. Quantum mechanics has clarified much for physical chemistry by modeling the smallest particles ordinarily dealt with in the field, atoms and molecules, enabling theoretical chemists to use computers and sophisticated mathematical techniques to understand the chemical behaviour of matter. Chemical thermodynamics deals with the relationship between heat and other forms of chemical energy, kinetics with chemical reaction rates. Sub disciplines of physical chemistry include electrochemistry, photochemistry, surface chemistry, and catalysis.

Electrochemistry

Electrochemistry is the sub discipline of chemistry that deals with the study of the relationship between electrical energy and chemical changes. Chemical reactions that involve the input or generation of electric currents are called electrochemical reactions. Such reactions are broadly classified into two categories:

- Production of chemical change by electrical energy i.e. the phenomenon of electrolysis.
- Conversion of chemical energy into electrical energy. i.e., the generation of electricity by spontaneous redox reactions.

Electricity can be produced when electrons move from one element to another in certain types of reactions (such as redox reactions). Typically, electrochemistry deals with the overall reactions when multiple redox reactions occur simultaneously, connected via some external electric current and a suitable electrolyte. In other words, electrochemistry is also concerned with chemical phenomena that involve charge separation (as seen commonly in liquids such as solutions). The dissociation of charge often involves charge transfer that occurs homogeneously or heterogeneously between different chemical species.

Electrochemical Cell

A spontaneous chemical process is the one which can take place on its own and in such a process Gibbs free energy of a system decreases. In electrochemistry, spontaneous reaction (redox reaction) results in the conversion of chemical energy into electrical energy. The reverse process is also possible where a non-spontaneous chemical reaction occurs by supplying electricity. These interconversions are carried out in equipment called electrochemical cell.

Types of Electrochemical Cell

Electrochemical cell are of two types: Galvanic cells and electrolytic cells.

Galvanic Cell

The galvanic cell converts chemical energy into electrical energy i.e. electricity can be obtained with the help of redox reaction. The oxidation and reduction take place in two separate compartments. Each compartment consists of an electrolyte solution and metallic conductor which acts as an electrode. The compartment containing the electrode and the solution of the electrolyte is called half cells. For example: Daniell cell is a galvanic cell in which zinc and copper are used for a redox reaction to take place.

$$Zn(s) + Cu^{+2}(aq) \rightarrow Zn^{+2} + Cu(s)$$

- At anode (oxidation half): $Zn(s) \rightarrow Zn^{+2} + 2e^-$.
- At cathode(reduction half): $Cu^{+2}(aq) + 2e^- \rightarrow Cu(s)$.

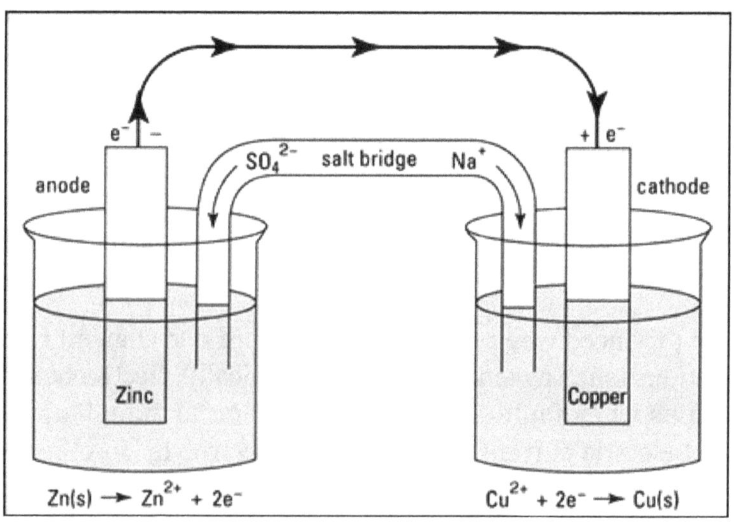

Salt Bridge

Salt bridge is usually an inverted U-tube filled with a concentrated solution of inert electrolytes. It is used to maintain the charge balance and to complete the circuit by allowing the flow of ions through it. It contains a gel in which inert electrolytes like KNO_3 or K_2SO_4 are mixed. Through the salt bridge, negative ion flows towards the anode and positive ion flow to the cathode and the charge balance is maintained and cell keeps on functioning.

Electrode Potential

In a galvanic cell, when two-electrode are dipped in their respective ion there is a tendency of one of the electrodes (anode) to undergo oxidation whereas the ion of the other electrode (cathode) has the tendency to gain an electron. This tendency of losing of electrons (oxidation) or gaining of electrons (reduction) is called electrode potential.

Standard Electrode Potential (E°)

Standard electrode potential is defined as the electrode potential of an electrode relative to standard hydrogen electrode under standard conditions. The standard conditions taken are:

- 1 molar concentration of each ion in the solution.
- A Temperature of 298 K.
- 1 bar pressure.

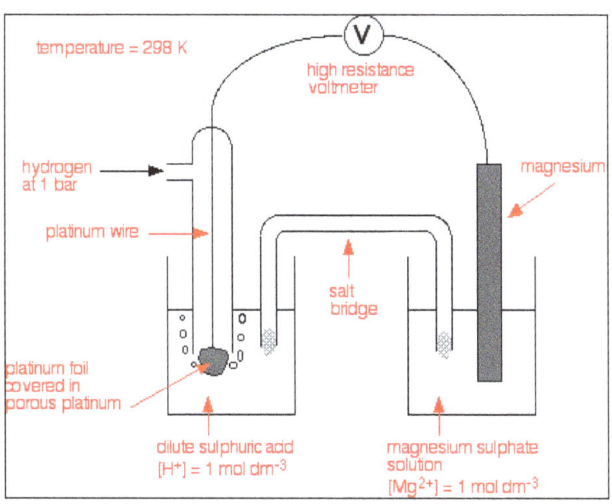

Electrochemical Series

A series of the standard electrode has been established by measuring the potential of various electrodes versus standard hydrogen electrode (SHE). When the electrodes (metals and non-metals) in contact with their ions are arranged on the basis of the values of their standard reduction potential or standard oxidation potentials, the resulting series is called the electrochemical series.

Elements Reduction Reaction

Cell Potential or EMF of a Cell

$F_2(g) + 2e^-$	→	$2F^-(aq)$	+2.87
$Au^+(aq) + e^-$	→	$Au(s)$	+1.68
$Cl_2(g) + 2e^-$	→	$2Cl^-(aq)$	+1.36
$O_2(g) + 4H^+(aq) + 4e^-$	→	$2H_2O(1)$	+1.23
$Ag^+(aq) + e^-$	→	$Ag(s)$	+0.80
$Fe^{3+}(aq) + e^-$	→	$Fe^{2+}(aq)$	+0.77
$I_2(s) + 2e^-$	→	$2I^-(aq)$	+0.54
$O_2(g) + 2H_2O(l) + 4e^-$	→	$4OH^-(aq)$	+0.40
$Cu^{2+}(aq) + 2e^-$	→	$Cu(s)$	+0.34
$2H^+(aq) + 2e^-$	→	$H_2(g)$	0.00
$Pb^{2+}(aq) + 2e^-$	→	$Pb(s)$	−0.13
$Sn^{2+}(aq) + 2e^-$	→	$Sn(s)$	−0.14
$Ni^{2+}(aq) + 2e^-$	→	$Ni(s)$	−0.23
$Co^{2+}(aq) + 2e^-$	→	$Co(s)$	−0.28
$Fe^{2+}(aq) + 2e^-$	→	$Fe(s)$	−0.44
$Zn^{2+}(aq) + 2e^-$	→	$Zn(s)$	−0.76
$2H_2O(l) + 2e^-$	→	$H_2(g) + 2OH^-(aq)$	−0.83
$Al^{3+}(aq) + 3e^-$	→	$Al(s)$	−1.67
$Mg^{2+}(aq) + 2e^-$	→	$Mg(s)$	−2.34
$Na^+(aq) + e^-$	→	$Na(s)$	−2.71
$Ca^{2+}(aq) + 2e^-$	→	$Ca(s)$	−2.87
$K^+(aq) + e^-$	→	$K(s)$	−2.93

In the galvanic cell there are two half-cell, the oxidation half-cell (anode) and the reduction half-cell (cathode).Due to the difference in the potentials of these half-cells, the electric current moves from the electrode of higher potential (cathode) to the electrode of lower potential (anode). The difference between the electrode potential of the two half-cell is called cell potential or emf of a cell.

- $E^o_{cell} = E_{Cathode} - E_{anode}$.
- $E_{Cathode}$ = Standard reduction potential of the cathode.
- E_{anode} = Standard reduction potential of the anode.
- If E^o_{cell} is positive then the reaction is feasible.
- If E^o_{cell} is negative then the reaction is not feasible.

Electrode and Cell Potentials-Nernst Equation

The electrode potential and the emf of the cell depend upon the nature of the electrode, temperature and the activities (concentrations) of the ions in solution. For the general electrochemical reaction of the type:

$$aA + bB \rightarrow cC + dD$$

$$E_{cell} = E^{o}_{cell} - RT/nF \ln [C]^{c}[D]^{d}/[A]^{a}[B]^{b}$$

$$E_{cell} = E^{o}_{cell} - 2.0303\ RT/nF \log [C]^{c}[D]^{d}/[A]^{a}[B]^{b}$$

$$E_{cell} = E^{o}_{cell} - 0.0591/n \log [C]^{c}[D]^{d}/[A]^{a}[B]^{b} \text{ at } 298K$$

Electrolytic Cell

The electrolytic cell converts electrical energy to chemical energy. Here the electrodes are dipped in an electrolytic solution containing cations and anions. On supplying current the ions move towards electrodes of opposite polarity and simultaneous reduction and oxidation take place.

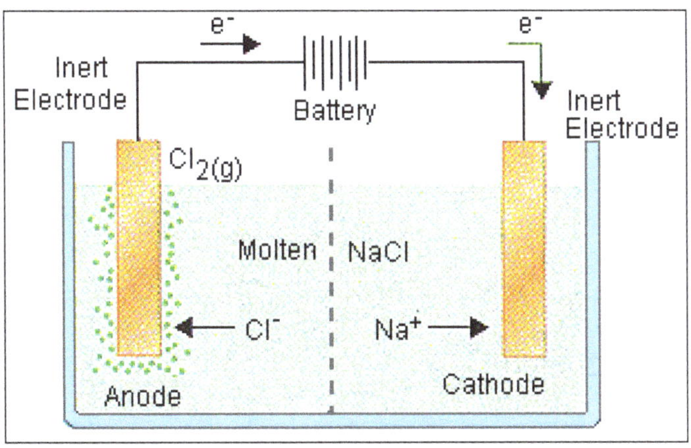

For example, in the electrolysis of molten sodium chloride, sodium chloride is melted (above 801 °C), two electrodes are inserted into the melt, and an electric current is passed through the molten salt. The chemical reaction that takes place at the electrodes are:

- Sodium-ion migrates to the cathode, where sodium ion gain one electron and reduce to sodium metal.

  $$Na^{+} + e^{-} \rightarrow Na$$

- Chloride ions migrate towards the anode where it loses one electron and gets oxidised to chlorine gas.

  $$Cl^{-} \rightarrow 1/2\ Cl_2 + e^{-}$$

- The overall reaction is the breakdown of sodium chloride into its elements:

  $$2NaCl \rightarrow 2Na(s) + Cl_2(g)$$

Preferential Discharge of Ions

When there are more than one cations or anion the process of discharge becomes competitive in nature. For example, in the electrolysis of NaCl solution, apart from Na^+ and Cl^- ions the solution of sodium chloride also contains H^+ and OH^- ions due to ionisation of water. When the potential difference is applied between the two electrodes, Na^+ and H^+ ions move towards the cathode and Cl^- and OH^- ions move towards the anode. At cathode H^+ ions get reduce in preference to giving hydrogen gas because hydrogen has a higher reduction potential than sodium. Similarly, at the anode, Cl^- ions are oxidised in preference to OH^- to give chlorine gas.

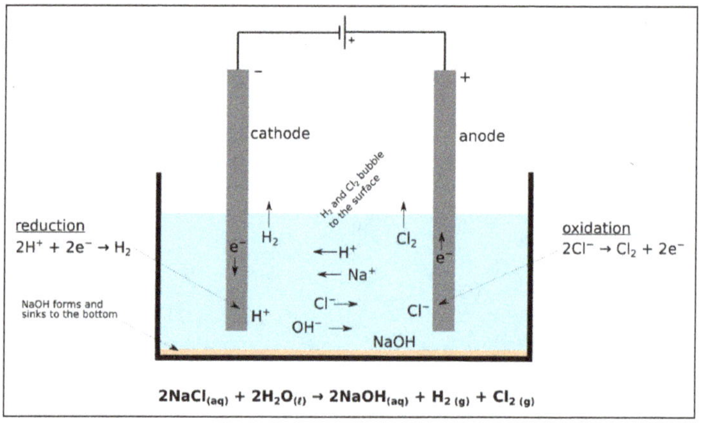

Faraday's Law of Electrolysis

The relationship between the quantity of electric charge passed through an electrolyte and the amount of substance deposited at the electrodes was given by Faraday in 1834, in the form of the law of electrolysis.

Faraday's First Law

When an electric current is passed through an electrolyte, the amount of substance deposited is proportional to the quantity of electric charge passed through the electrolyte. If W is the mass of the substance deposited by passing Q coulomb of charge then according to this law:

$$W \propto Q$$

Now, $Q = I \times t$

$$W \propto I \times t$$

$$W = z \times I \times t$$

where Z is a constant known as electrochemical equivalent and is characteristic of a substance deposited.

Faraday's Constant(F)

It is the charge possessed by 1 mole of electrons and it is equal to 96500 coulombs(approx.). In terms of Faraday's constant the number of gram equivalent of electrolyte discharged at an electrode is equal to the faraday's passed,

$$W = E \times Q / 96500$$

Faraday's Second Law

When the same quantity of charge is passed through different electrolytes, then the mass of different substance deposited at the respective electrodes will be in the ratio of their equivalent masses. Mathematically it is represented as:

$$W_1/W_2 = Z_1/Z_2$$

Where W_1 and W_2 are the weight of two substances which are deposited at their respective electrodes and Z_1 and Z_2 are their respective equivalent weight.

Examples: Given that:

$$Fe^{2+}/Fe = -0.44\ v,\ Ni^{+2}/Ni = 0.25\ v$$

$$Ag^{+}/Ag = 0.80\ v,\ Cu^{+2}/Cu = 0.34\ v$$

Which of the following reactions under standard condition will not take place in the specified direction.

$$Ni^{+2}(aq) + Cu(s) \rightarrow Ni(s) + Cu^{+2}(aq)$$
$$Cu(s) + 2\,Ag^{+}(aq) \rightarrow Cu^{+2}(aq) + 2\,Ag(s)$$
$$Cu(s) + 2H^{+}(aq) \rightarrow Cu^{+2} + H_2(g)$$
$$Fe(s) + 2H^{+}(aq) \rightarrow Fe^{+2}(aq) + 3\,H_2$$

Solution: Option a and c.

$$Ni^{+2}(aq) + Cu(s) \rightarrow Ni(s) + Cu^{+2}(aq)$$
$$E^{o}_{cell} = 0.25 - 0.34$$
$$= -0.11V$$

Reaction is not feasible (negative E^{o}_{cell}):

$$Cu(s) + 2\,Ag^{+}(aq) \rightarrow Cu^{+2}(aq) + 2\,Ag(s)$$
$$E^{o}_{cell} = 0.80 - 0.34 = 0.46\ v$$

Reaction is feasible (positive E^o_{cell}):

$$Cu(s)+2H^+(aq)\rightarrow Cu^{+2}+H_2(g)$$
$$E^o_{cell}=0-0.34=-0.34\ v$$

Reaction is not feasible (negative E^o_{cell}):

$$Fe(s)+2H^+(aq)\rightarrow Fe^{+2}(aq)+3\ H_2$$
$$E^o_{cell}=0-(-0.44)=0.44\ v$$

Reaction is feasible (positive E^o_{cell}).

Example: The emf of a cell corresponding to the reaction, Zn + 2H⁺ (aq) → Zn^{+2} (0.1M) + H_2(g) 1 atm is 0.30 v at 25 °C. Write the half-cell reactions and calculate the pH of the solution at the hydrogen electrode. (E^o_{cell} = − 0.76 v).

Solution:

$$E^o_{cell}=0-(-0.76)=0.76\ v$$

Applying the Nernst equation,

$$E_{cell}=E^o_{cell}-0.0591/2\log\left[zN^{+2}\right][H_2]/[H]^+$$
$$0.30=0.76-0.0591/2\log 0.1\times 1/\left[H^+\right]^2$$
$$\log 0.1/\left[H^+\right]^2=2\times 0.46/0.0591$$
$$\log 0.1-\log\left[H^+\right]^2=15.56$$
$$2\ pH=15.56-\log 0.1$$
$$pH=16.56/2=8.28$$

Example: The solution of metal of atomic mass X was electrolyzed for 1 hour with a current of 0.25 ampere. The mass of metal deposited was 0.295g. Find the metal X if its valency is 2.

Solution: Given, I = 0.25 ampere, t= 1 hr = 60 × 60 = 3600s

$$Q=I\times t$$
$$Q=0.25\times 3600$$
$$=900\ coulombs$$

Therefore, 900 coulombs of electricity deposit = 0.295.

96500 coulomb of electricity deposit = 0.295 × 96500/900 = 31.63g.

Valency of metal = atomic mass/equivalent mass.

Atomic mass of metal X = 31.63 × 2 = 63.26g.

Therefore, the metal X is copper.

Chemical Kinetics

Chemical kinetics is the branch of physical chemistry that is concerned with understanding the rates of chemical reactions. It is to be contrasted with thermodynamics, which deals with the direction in which a process occurs but in itself tells nothing about its rate. Thermodynamics is time's arrow, while chemical kinetics is time's clock. Chemical kinetics relates to many aspects of cosmology, geology, biology, engineering, and even psychology and thus has far-reaching implications. The principles of chemical kinetics apply to purely physical processes as well as to chemical reactions.

One reason for the importance of kinetics is that it provides evidence for the mechanisms of chemical processes. Besides being of intrinsic scientific interest, knowledge of reaction mechanisms is of practical use in deciding what is the most effective way of causing a reaction to occur. Many commercial processes can take place by alternative reaction paths, and knowledge of the mechanisms makes it possible to choose reaction conditions that favour one path over others.

A chemical reaction is, by definition, one in which chemical substances are transformed into other substances, which means that chemical bonds are broken and formed so that there are changes in the relative positions of atoms in molecules. At the same time, there are shifts in the arrangements of the electrons that form the chemical bonds. A description of a reaction mechanism must therefore deal with the movements and speeds of atoms and electrons. The detailed mechanism by which a chemical process occurs is referred to as the reaction path, or pathway.

The vast amount of work done in chemical kinetics has led to the conclusion that some chemical reactions go in a single step; these are known as elementary reactions. Other reactions go in more than one step and are said to be stepwise, composite, or complex. Measurements of the rates of chemical reactions over a range of conditions can show whether a reaction proceeds by one or more steps. If a reaction is stepwise, kinetic measurements provide evidence for the mechanism of the individual elementary steps. Information about reaction mechanisms is also provided by certain nonkinetic studies, but little can be known about a mechanism until its kinetics has been investigated. Even then, some doubt must always remain about a reaction mechanism. An investigation, kinetic or otherwise, can disprove a mechanism but can never establish it with absolute certainty.

Reaction Rate

The rate of a reaction is defined in terms of the rates with which the products are formed and the reactants (the reacting substances) are consumed. For chemical systems it is usual to deal with the concentrations of substances, which is defined as the amount of substance per unit volume. The rate can then be defined as the concentration of a substance that is consumed or produced in unit time. Sometimes it is more convenient to express rates as numbers of molecules formed or consumed in unit time.

The Half-Life

A useful rate measure is the half-life of a reactant, which is defined as the time that it takes for half of the initial amount to undergo reaction. For a special type of kinetic behaviour (first-order kinetics; see below Some kinetic principles), the half-life is independent of the initial amount. A common and straightforward example of a half-life independent of the initial amount is radioactive substances. Uranium-238, for example, decays with a half-life of 4.5 billion years; of an initial amount of uranium, half of that amount will have decayed in that period of time. The same behaviour is found in many chemical reactions.

Even when the half-life of a reaction varies with the initial conditions, it is often convenient to quote a half-life, bearing in mind that it applies only to the particular initial conditions. Consider, for example, the reaction in which hydrogen and oxygen gases combine to form water; the chemical equation is,

$$2H_2 + O_2 \rightarrow 2H_2O.$$

If the gases are mixed together at atmospheric pressure and room temperature, nothing observable will happen over long periods of time. However, reaction does occur, with a half-life that is estimated to be more than 12 billion years, which is roughly the age of the universe. If a spark is passed through the system, the reaction occurs with explosive violence, with a half-life of less than one-millionth of a second. This is a striking example of the great range of rates with which chemical kinetics is concerned. There are many possible processes that proceed too slowly to be studied experimentally, but sometimes they can be accelerated, often by the addition of a substance known as a catalyst. Some reactions are even faster than the hydrogen-oxygen explosion—for example, the combination of atoms or molecular fragments (called free radicals) where all that occurs is the formation of a chemical bond. Some modern kinetic investigations are concerned with even faster processes, such as the breakdown of highly energetic and therefore transient molecules, where times of the order of femtoseconds (fs; 1 fs = 10^{-15} second) are involved.

Measuring Slow Reactions

The best way to study exceedingly slow reactions is to change the conditions so that the reactions occur in a reasonable time. Increasing the temperature, which can have a strong effect on the reaction rate, is one possibility. If the temperature of a hydrogen-oxygen mixture is raised to about 500 °C (900 °F), reaction then occurs rapidly, and its kinetics has been studied under those conditions. When a reaction occurs to a measurable extent over a period of minutes, hours, or days, rate measurements are straightforward. Amounts of reactants or products are measured at various times, and the rates are readily calculated from the results. Many automated systems have now been devised for measuring rates in this way.

Measuring Fast Reactions

Some processes are so fast that special techniques have to be used to study them. There are two difficulties with fast reactions. One is that the time that it takes to mix reactants or to change the temperature of the system may be significant in comparison with the half-life, so that the initial time cannot be measured accurately. The other difficulty is that the time it takes to measure the amounts of substances may be comparable with the half-life of the reaction. The methods used to overcome these difficulties fall into two classes: flow methods and pulse and probe methods.

In flow methods, two gases or solutions are introduced rapidly into a mixing vessel, and the resulting mixture then flows rapidly along a tube. Concentrations of reactants or products may then be measured—for example, by spectroscopic methods—at various positions along the tube, which correspond to various reaction times. A modification of this method is the stopped-flow technique, in which the reactants are forced rapidly into a reaction chamber; the flow is then suddenly stopped, and the amounts are measured by physical methods after various short times. These flow methods are limited by the time it takes to mix gases or solutions and are not suitable if the half-life is less than about a hundredth of a second.

These mixing difficulties were overcome by pulse and probe methods. The principle of these is that a short pulse, usually of radiation, is given to a chemical system and is then followed by a probe, usually involving radiation that provides spectroscopic evidence of what occurred after the initial pulse. The first of these methods, developed in 1949 by British chemists R.G.W. Norrish and George Porter, was the flash-photolysis method, for which Norrish and Porter won the Nobel Prize for Chemistry in 1967. In this technique a flash of light of high intensity but short duration brings about the formation of atomic and molecular species, the reactions of which can be studied kinetically by spectroscopy. In the earliest experiments the duration of the flash was about a millisecond (ms; 1 ms = 10^{-3} second), but in the next four decades the duration was reduced by more than 11 powers of 10, to just a few femtoseconds. A nanosecond (ns; 1 ns = 10^{-9} second) flash is adequate for studying almost any purely chemical reaction where there is a change in chemical identity. Any chemical reaction, however, involves processes of a purely physical nature, such as energy redistribution and the breakdown of transient species, which occur in the femtosecond range.

Many such processes have now been studied with flashes of only a few femtoseconds' duration. The time that it takes for the length of a chemical bond to change by 10^{-10} metre can be as little as about 100 fs, so that a flash of a few femtoseconds' duration, closely followed by another one of the same duration, will provide information about such tiny changes in bond lengths. The technique for causing one flash to occur a few nanoseconds after another is to route the light by a slightly longer path. A path of 1 additional micrometre (μm; 1 μm = 10^{-6} metre) causes a delay of 1 fs, and such a short path difference is now technically feasible. Egyptian-born chemist Ahmed Zewail won the Nobel Prize for Chemistry in 1999 for his work in this field.

Another pulse method is the relaxation method, developed in the 1950s by German physicist Manfred Eigen (who shared the Nobel Prize for Chemistry in 1967 with Norrish and Porter). In this method the investigation begins with a reaction system in equilibrium; the reaction to be studied has finished, and no further changes take place. The external conditions are then altered very rapidly; the system is then no longer at equilibrium, and it relaxes to a new equilibrium. The speed of relaxation is measured by a physical method such as spectroscopy, and analysis of the results leads to the reaction rate.

The most common way of changing the external conditions is to change the temperature, and the method is called the temperature-jump, or T-jump, method. Techniques have been developed for raising the temperature of a tiny reaction vessel by a few degrees in less than 100 ns. The method is therefore not suitable for the fastest processes, which can be studied by flash photolysis, but many purely chemical processes are suitable for the T-jump technique, which has provided valuable kinetic information.

Other experimental techniques are used for the study of rapid processes. Ultrasonic methods have been used for processes occurring with half-lives in the microsecond (μs; 1 μs = 10^{-6} second) and nanosecond ranges. Nuclear magnetic resonance has also been used for certain types of reactions.

Kinetic Principles

The kinetic behaviour of an ordinary chemical reaction is conventionally studied in the first instance by determining how the reaction rate is influenced by certain external factors such as the concentrations of the reacting substances, the temperature, and sometimes the pressure. For a reaction in which two substances A and B react with each other, it is sometimes found that the reaction rate is proportional to the concentration of A, represented by [A], and to the concentration of B, or [B]. In that case the reaction is said to be a second-order reaction; it is first order in [A] and first order in [B]. In such a case the reaction rate υ can be expressed as:

$$\upsilon = k[A][B],$$

where k is a constant, known as the rate constant for the reaction.

This is just one of many types of kinetics that can be observed. A substance A that changes into another substance may obey a kinetic equation of the form $v = k[A]$, which is a first-order reaction. It is important to recognize that the kinetics of a reaction does not always correspond in a simple way to the balanced chemical equation for the reaction. Thus, if a reaction is of the form,

$$A + B \rightleftharpoons Y + Z,$$

the reaction is not necessarily second-order in both directions. This is in contrast to the situation with the equilibrium constant for the reaction, which corresponds to the balanced equation. The reason why the kinetic law is different is that the reactions in the forward and reverse directions may occur by stepwise mechanisms that lead to a different and usually more complex kinetic equation.

Sometimes reaction rates depend on reactant concentrations in a more complicated way. This is a clear indication that a reaction happens in several steps.

The effect of temperature on reaction rates provides much information about reaction mechanisms. Understanding of this effect owes much to the ideas of the Dutch physical chemist Jacobus Henricus van 't Hoff and the Swedish chemist Svante August Arrhenius. Their equation for the dependence of a rate constant k on the absolute temperature T is,

$$k = A \exp(-E/RT),$$

where R is the molar gas constant and A and E are quantities that are different for each reaction. This equation has come to be called the Arrhenius equation, although, as Arrhenius acknowledged when he applied it in 1889, it was first suggested by van't Hoff in 1884. According to this relationship, a plot of the logarithm of the rate constant against the reciprocal of the absolute temperature should yield a straight line. From the slope and intercepts of the line, it is possible to calculate the value of the kinetic parameters A and E. The Arrhenius relationship applies satisfactorily to most reactions and indeed to many physical processes; however, various complications may cause it to fail.

If the reaction between two molecules is an elementary one, occurring in a single step, a simple interpretation of the Arrhenius equation can be given. The quantity A is related to the frequency of collisions between the reacting molecules. The quantity E, known as the activation energy for the reaction, results from the fact that there is an energy barrier to reaction. If E was zero, k would be equal to A, which means that the reaction would occur every time a collision occurred between the reactant molecules. This is the case for reactions in which no chemical bond is broken, such as the combination of atoms.

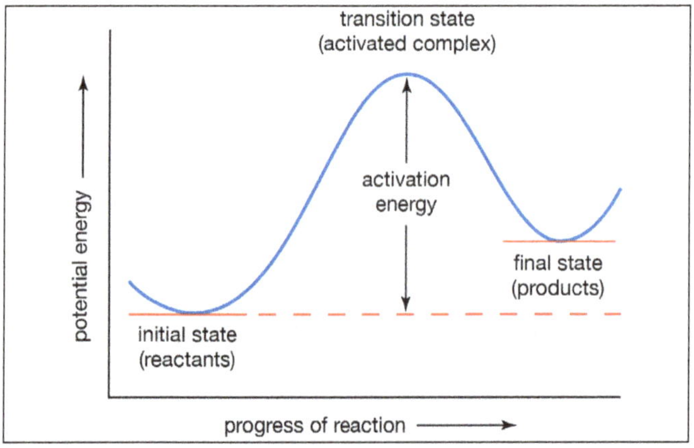

Activation energy

In the Arrhenius equation, the activation energy (E) represents the minimum amount of energy required to transform reactants into products in a chemical reaction. On a potential energy curve, the value of the activation energy is equivalent to the difference in potential energy between particles in an intermediate configuration (known as the activated complex, or transition state) and particles of reactants in their initial state. The activation energy thus can be visualized as a barrier that must be overcome by reactants before products can be formed.

For reactions in which a chemical bond is broken, on the other hand, the activation energy E is not zero but has a value that is often a tenth or so of the energy required to break the bond. A simple and essentially correct explanation of the activation energy was suggested by Arrhenius, who pointed out that, for many reactions, raising the temperature by 10 °C (18 °F) doubles the reaction rate. This increase cannot be caused by the increase in the frequency of collisions between colliding molecules, since the frequency does not increase sufficiently with a rise in temperature. Arrhenius suggested that when reactants A and B react together, they first form a highly energized intermediate that is denoted as $AB^{\ddagger}$, which subsequently gives the products of reaction:

$$A + B \rightleftharpoons AB^{\ddagger} \rightarrow Y + Z.$$

If the intermediate complex (also called the activated complex) $AB^{\ddagger}$ is of high energy, it is formed only in small amounts. According to the Boltzmann principle, the fraction of molecules having energy greater than E is $\exp(-E/RT)$, which provides an explanation of the appearance of this fraction in the Arrhenius equation. The interpretation of the equation is thus that only those molecules having energy greater than E are able to undergo reaction; other collisions are ineffective, and the reactant molecules merely separate unchanged.

Composite Reaction Mechanisms

Various lines of evidence are used to determine if a reaction occurs in more than one step. Suppose that the kinetic equation for the reaction does not correspond to the balanced equation for the reaction. A simple example is the reaction between hydrogen and iodine chloride, with the formation of iodine and hydrogen chloride:

$$H_2 + 2ICl \rightarrow I_2 + 2HCl.$$

To make the equation balance, the reaction must be written as shown, with two iodine chloride molecules reacting with a single hydrogen molecule. If this reaction occurred in a single elementary step, the rate would be proportional to the first power of the hydrogen concentration and the square of the iodine chloride concentration. Instead, however, the rate is found to be proportional to both concentrations to the first power, so that it is a second-order reaction:

$$\upsilon = k[H_2][ICl].$$

This can be explained if there is initially a slow reaction between one hydrogen molecule and one of iodine chloride:

$$H_2 + ICl \rightarrow HI + HCl \ (\text{slow})$$

Followed by a rapid reaction between the hydrogen iodide formed and an additional molecule of iodine chloride:

$$HI + ICl \rightarrow HCl + I_2 \ (\text{fast}).$$

If the second reaction is fast, the hydrogen iodide is removed as fast as it is formed. The rate of the second reaction therefore has no effect on the overall rate, which is the rate of the first step. This mechanism therefore explains the kinetic behaviour but does not prove it; other, more complicated schemes could be devised, but, until there is further evidence, it is expedient to accept the simple mechanism. This is an example of a consecutive reaction, which occurs in two steps, with the intermediate playing a role.

Another piece of evidence for a composite mechanism is the detection of reaction intermediates. In such a case, a reaction scheme must be devised that will account for these intermediates. Sometimes an intermediate can be a fairly stable substance. In other cases the intermediates are unstable species such as atoms and free radicals (fragments of molecules) that subsequently undergo rapid reactions. Free radicals can be detected by spectroscopy and other means. When organic molecules are raised to high temperatures, they decompose into smaller molecules, and organic free radicals have often been detected as intermediates. In an explosion, such as that between hydrogen and oxygen, free radicals such as hydroxyl can be detected.

Composite reaction mechanisms are of various kinds. Aside from the simple consecutive schemes, there are some special mechanisms that give rise to oscillatory behaviour: the amount of a product continuously rises and falls over a period of time. The conditions for this behaviour are that there must be at least two species involved in the reaction and there must be feedback, which means that products of the reaction affect the rate. There are also reaction mechanisms that give rise to what is technically known as chaos, or catastrophe. With such reactions it is impossible to predict the outcome. Chaotic conditions also require that there be feedback and that at least three species be involved.

Sometimes a complex reaction mechanism involves a cycle of reactions such that certain intermediates consumed in one step are regenerated in another. For example, the accepted mechanism of the reaction between hydrogen and bromine, which can be written as:

$$H_2 + Br_2 \rightarrow 2HBr,$$

Includes the steps:

$$Br + H_2 \rightarrow HBr + H$$
$$H + Br_2 \rightarrow HBr + Br.$$

In the first of these steps a bromine atom is consumed, but in the second a bromine atom is regenerated. This pair of reactions can thus occur with the production of two molecules of hydrogen bromide, the product of the reaction, without loss of bromine atoms. This pair of reactions is called a cycle of reactions, and it can occur a number of times, in which case the reaction is referred to as a chain reaction. The two reactions in which bromine is regenerated are known as the chain-propagating steps. The average number of times the pair of steps is repeated is known as the chain length.

One necessary condition for a proposed reaction mechanism to be correct is that it must account for the overall kinetic behaviour of the reaction—in particular, for the dependence of the reaction rate on the reactant concentrations. For any proposed reaction mechanism, it is possible to write down equations for the rate of each step in terms of the reactant concentration and then to solve the equations for the overall rate. A practical difficulty arises, since no exact mathematical solution is possible for all except the simplest of mechanisms. If one has values for the rate constants, solutions can be obtained with a computer, but explicit rate equations provide more insight into the reactions. One therefore looks for approximate solutions of the equations. One of these is provided by the steady-state treatment, which is applicable if (and only if) the intermediates are species that can be present only at low concentrations. If this condition is satisfied by an intermediate, the rate of change of its concentration during the course of reaction is always small and, as a good approximation, can be assumed to be zero, which means that the intermediate exists in a steady state. This approximation may safely be applied to atoms and free radicals present as reaction intermediates. With this approximation it is usually possible to obtain a reliable approximate equation for the overall reaction rate in terms of reactant concentrations. If this agrees with the experimental behaviour, the mechanism is accepted.

One situation to which the steady-state treatment does not apply is when a reaction is an explosion. Explosions occur because the concentration of intermediates does not remain steady during the course of reaction but rises to a high value so that the reaction goes out of control. This occurs if the reaction mechanism involves a special kind of chain called a branching chain. In the hydrogen-oxygen explosion, for example, the following reaction is known to occur:

$$H + O_2 \rightarrow OH + O.$$

In this step a single chain carrier hydrogen atom has produced two chain carriers: a hydroxyl group and an oxygen atom. The number of chain carriers increases rapidly and leads to an explosion.

Theories of Reaction Rates

Two different theoretical approaches to chemical kinetics have led to an understanding of the details of how elementary chemical reactions occur. Both of these are based on the idea of potential-energy surfaces, which are models showing how the potential energy of a reaction system varies with certain critical interatomic distances. The course of an elementary reaction is represented by the movement of the system over the potential-energy surface. One theoretical approach to the problem involves studying the region of the potential-energy surface that corresponds to the highest point on the energy barrier that separates the reactants from the products. This approach is relatively simple and leads to explicit general expressions for the reaction rate. The second approach involves considering the dynamics of the motion of the system over the potential-energy surface.

Transition-State Theory

The idea of a potential-energy surface sprang from the ideas of Dutch physical chemist Jacobus Henricus van 't Hoff and Swedish physicist Svante August Arrhenius that were put forward to explain the effect of temperature on reaction rates. An important advance was made in 1931 by American chemist Henry Eyring and British chemist Michael Polanyi, who constructed, on the basis of quantum mechanics, a potential-energy surface for the simple reaction:

$$H^\alpha + H^\beta - H^\gamma \rightarrow H^\alpha - H^\beta - H^\gamma \rightarrow H^\alpha - H^\beta + H^\gamma.$$

For convenience the labels α, β, and γ are added as superscripts. When this reaction occurs, an atom H^α attacks a hydrogen molecule $H^\beta{-}H^\gamma$ and abstracts one of the hydrogen atoms from it. As the bond begins to form, the $H^{\beta}-H^{\gamma}$ bond becomes more and more extended and finally breaks. Somewhere along the reaction path, there is a particular intermediate state corresponding to the maximum value of the potential energy.

This particular intermediate state is usually designated by the superscript ‡ (used above in the discussion of the Arrhenius equation). It is known as an activated complex and plays an important role in what has come to be called transition-state theory, developed independently in 1935 by Eyring, Polanyi, and English physical chemist M.G. Evans. The essential feature of the theory is that the activated complexes are considered to be formed from the reactants in a state in which they are in equilibrium with the reactants. Thus, the above reaction can be written as

$$H^{\alpha} + H^{\beta} - H^{\gamma} \rightarrow H^{\alpha} - H^{\beta} - H^{\gamma\ddagger}(\text{at equilibrium with } H^{\alpha} + H^{\beta} - H^{\gamma}) \rightarrow H^{\alpha} - H^{\beta} + H^{\gamma}.$$

Since the activated complexes are in equilibrium, their concentration can be expressed in terms of the concentrations of the reactants. The reaction rate is this concentration multiplied by the frequency with which they form products, which is known from kinetic theory. Despite the approximations involved in transition-state theory, it has been successful in providing an insight into how chemical reactions occur and how their rates depend on various factors.

Molecular Dynamics

The second theoretical approach to chemical kinetics is referred to as molecular dynamics, or reaction dynamics. It is a more detailed treatment of reactions and is designed to investigate the atomic motions that occur during a chemical reaction and the quantum states of the reactant and product molecules. Such studies are important in testing the validity of transition-state theory and similar treatments. Also, there are important practical applications of kinetics, such as reactions occurring in lasers, for which information about the energy states of the products of a chemical reaction is needed; this information is not provided by transition-state theory but is an important outcome of molecular dynamics.

Consider a simple reaction of the type A + B−C → Y + Z, where A is an atom and B−C is a diatomic molecule. A dynamical calculation would first involve calculating, using quantum mechanics, a potential-energy surface that gives the potential energy corresponding to a set of initial configurations. One can make dynamical calculations for a variety of vibrational states of the reactant B−C and for a variety of translational energies. Ideally the calculations would be based on quantum mechanics, but this proves difficult, and often classical mechanics is used. Except for certain types of reactions where quantum effects are important, it appears that for many reaction systems the error involved in neglecting quantum effects is not great. In principle the dynamical calculations should give more-reliable results than any other treatment; however, for any but the simplest reactions, the computer calculations are time-consuming, and, since many approximations must be made to save computer time, there is often some uncertainty about the results.

Much work has been done along these lines, and on the whole the agreement with both experiment and transition-state theory is satisfactory. In addition, some calculations suggest important generalizations about reactions. For example, the form of the potential-energy surface greatly influences whether energy released in a reaction resides in the vibrations of the product molecules or in their kinetic energy of translation. Experimental studies of simple reactions, particularly some that involve the study of light emission in reactions (chemiluminescence) and that use narrow molecular beams, have also contributed to knowledge of chemical reactions.

Equilibrium Chemistry

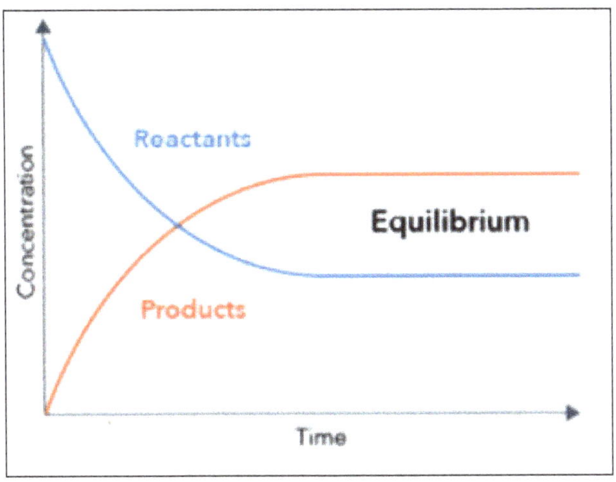

State of Chemical Equilibrium.

Chemical equilibrium refers to the state of a system in which the concentration of the reactant and the concentration of the products do not change with time and the system does not display any further change in properties.

When the rate of the forward reaction is equal to the rate of the reverse reaction, the state of chemical equilibrium is achieved by the system. When there is no further change in the concentrations of the reactants and the products due to the equal rates of the forward and reverse reactions, the system is said to be in a state of dynamic equilibrium.

A graph with the concentration on the y-axis and time on the x-axis can be plotted. Once the concentration of both the reactants and the products stops showing change, chemical equilibrium is achieved.

Why Chemical Equilibrium is Called Dynamic Equilibrium?

The stage at which the rate of the forward reaction is equal to the rate of backward reaction is called an equilibrium stage. At this point, the number of reactant molecules converting into products and product molecules into reactants are the same. The same equilibrium can be carried out with the same reactants anywhere with similar conditions with continuous interchanging of molecules hence chemical equilibrium is dynamic.

Types of Chemical Equilibrium

There are two types of chemical equilibrium:

- Homogeneous Equilibrium.
- Heterogeneous Equilibrium.

Homogenous Chemical Equilibrium

In this type, the reactants and the products of chemical equilibrium are all in the same phase. Homogenous equilibrium can be further divided into two types: Reactions in which the number of molecules of the products is equal to the number of molecules of the reactants. For example,

- $H_2\ (g) + I_2\ (g) \rightleftharpoons 2HI\ (g)$
- $N_2\ (g) + O_2\ (g) \rightleftharpoons 2NO\ (g)$

Reactions in which the number of molecules of the products is not equal to the total number of reactant molecules. For example,

- $2SO_2\ (g) + O_2\ (g) \rightleftharpoons 2SO_3\ (g)$
- $COCl_2\ (g) \rightleftharpoons CO\ (g) + Cl_2\ (g)$

Heterogeneous Chemical Equilibrium

In this type, the reactants and the products of chemical equilibrium are present in different phases. A few examples of heterogeneous equilibrium are listed below:

- $CO_2\ (g) + C\ (s) \rightleftharpoons 2CO\ (g)$
- $CaCO_3\ (s) \rightleftharpoons CaO\ (s) + CO_2\ (g)$

Thus, the different types of chemical equilibrium are based on the phase of the reactants and products.

Factors Affecting Chemical Equilibrium

According to Le-Chatelier's principle, if there is any change in the factors affecting the equilibrium conditions, the system will counteract or reduce the effect of the overall transformation. This principle applies to both chemical and physical equilibrium.

There are several factors like temperature, pressure and concentration of the system which affect equilibrium.

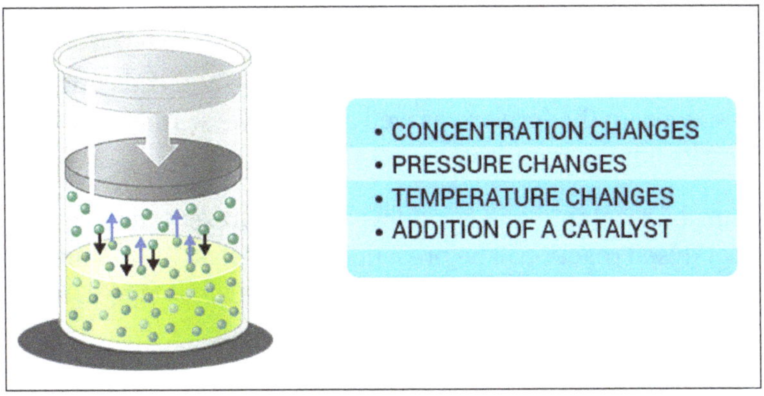

Change in Concentration

- The concentration of the reactants or products added is relieved by the reaction which consumes the substance which is added.
- The concentration of reactants or products removed is relieved by the reaction which is in the direction that replenishes the substance which is removed.
- When the concentration of the reactant or product is changed, there is a change in the composition of the mixture in chemical equilibrium.

Change in Pressure

Change in pressure happens due to the change in the volume. If there is a change in pressure it can affect the gaseous reaction as the total number of gaseous reactants and products are now different. According to Le Chatelier's principle, in heterogeneous chemical equilibrium, the change of pressure in both liquids and solids can be ignored because the volume is independent of pressure.

Change in Temperature

The effect of temperature on chemical equilibrium depends upon the sign of ΔH of the reaction and follows Le-Chatelier's Principle.

- As temperature increases the equilibrium constant of an exothermic reaction decreases.
- In an endothermic reaction the equilibrium constant increases with increase in temperature.

Along with equilibrium constant, the rate of reaction is also affected by the change in temperature. As per Le Chatelier's principle, the equilibrium shifts towards the reactant side when the temperature increases in case of exothermic reactions, for endothermic reactions the equilibrium shifts towards the product side with an increase in temperature.

Effect of a Catalyst

A catalyst does not affect the chemical equilibrium. It only speeds up a reaction. In fact, catalyst equally speeds up the forward as well as the reverse reaction. This results in the reaction reaching its equilibrium faster.

The same amount of reactants and products will be present at equilibrium in a catalysed or a non-catalysed reaction. The presence of a catalyst only facilitates the reaction to proceed through a lower-energy transition state of reactants to products.

Effect of Addition of an Inert Gas

When an inert gas like argon is added to a constant volume it does not take part in the reaction so the equilibrium remains undisturbed. If the gas added is a reactant or product involved in the reaction then the reaction quotient will change.

Examples of Chemical Equilibrium

In chemical reactions, reactants are converted into products by the forward reaction and the products may be converted into the reactants by the backward reaction. The two states, reactants and products are different in composition.

After some time of the start of the reaction, the rate of the forward and the backward reactions may become equal. After this, the number of reactants converted will be formed again by the reverse reaction such that the concentration of reactants and products do not change any more. Hence, the reactants and products are in chemical equilibrium.

- $N_2O_4 \rightleftharpoons 2NO_2$
- $PCl_5 \rightleftharpoons PCl_3 + PCl_2$
- $N_2 + H_2 \rightleftharpoons 2NH_3$

Importance of Chemical Equilibrium

It is useful in many industrial processes like:

- Preparation of ammonia by Haber's process: In this nitrogen combines with hydrogen to form ammonia, the yield of ammonia is more at low temperature, high pressure and in the presence of iron as catalyst.
- Preparation of sulphuric acid by contacts process: In this process, the fundamental

reaction is the oxidation of sulphur dioxide into sulphur trioxide. This involves chemical equilibrium.

Examples on Chemical Equilibrium

1. The equilibrium constant K_p for the reaction N_2 (g) + $3H_2$ (g) ⇌ $2NH_3$ (g) is 1.6×10^{-4} atm^{-2} at 400 °C. What will be the equilibrium constant of the Chemical equilibrium at 500°C if the heat of the reaction at this temperature range is -25.14 kcal?

Solution:

Equilibrium constants at different temperature and heat of the reaction are related by the equation,

$$\log KP_2 = -25140/2.303 \times 2\,[773 - 673 / 773 \times 673] + \log 1.64 \times 10^{-4}$$

$$\log KP_2 = -4.835$$

$$KP_2 = 1.462 \times 10^{-5}\ atm^{-2}$$

2. Given the equation, N_2 (g) + $3H_2$ (aq) ⇌ $2NH_3$ (g), Find Q and determine which direction the reaction will shift in order to reach the state of chemical equilibrium.

Solution:

Given, $[N_2]$ = 0.04M, $[H_2]$ = 0.09M, and K = 0.040.

Since only nitrogen and hydrogen concentration is given, it can be assumed as the reactants and ammonia as the product. Since ammonia concentration is not given it can be assumed to be zero.

As q is the ratio of the relative concentration of products to reactants, here Q = 0.

Since K = 0.04 is larger than Q, nitrogen and hydrogen will combine to form product ammonia.

Thermodynamics

Thermodynamics deals with energy changes accompanying all types of physical and chemical processes/transformations (Not concerned with total Energy of a body) Energy may be in forms: work, heat, light, mechanical or electrical etc.

Main Objectives of Chemical Thermodynamics

- To predict the feasibility of a process (may be physical or chemical) under the given conditions of Temperature, Pressure and Concentration.

- To predict the extent of feasible process: Thermodynamics has given certain mathematical relationships which can be employed to know the extent to which a transformation can take place before attainment of equilibrium.
- To deduce some important laws of Physical Chemistry: Generalization of physical chemistry such as gas equation, Law of Chemical Equilibrium, Phase Rule, Distribution rule/law, Van Hoff's Law of dilute solutions, Raoult's law of lowering of Vapour Pressure.
- To deduce equations relating to some physical properties: Equations co-relating the variations of certain physical properties (e.g. Temperature, Pressure) with others have been derived. From these laws. From these equations we can predict the effect of change of variable factors e.g. Temperature, Pressure etc. on different processes.

Thermodynamic System and its Surroundings

Any portion of universe or any portion of matter chosen for thermodynamic to study the effect of various variables such as Temperature, Pressure, Volume etc. is called a system. The remaining portion of the universe excluded from the system which may be capable of exchanging energy and matter with it, is called its surroundings. The system is separated from its surroundings by real or imaginary boundaries.

Types of Systems

Depending on the interactions between system and its surroundings three kinds of systems:

1. Isolated System: A system which can exchange neither energy nor matter with its surroundings is called an isolated system.

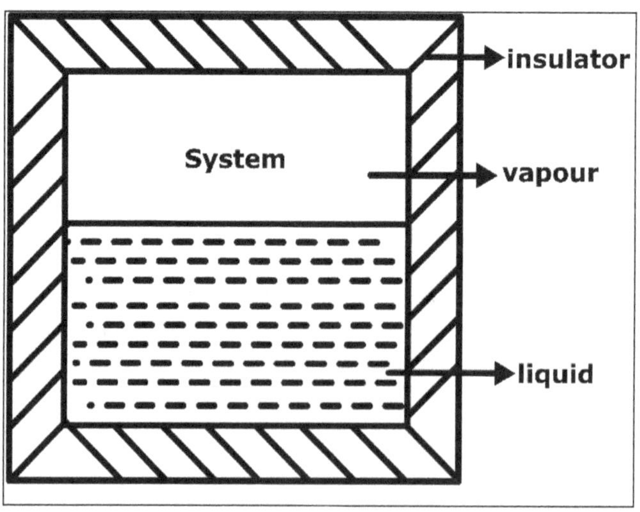

A liquid in contact with its vapor in an isolated closed vessel is an isolated system.

As the vessel is closed, vapor liquid can neither enter nor leave the vessel i.e. no exchange of matter from within the vessel with the surrounding. Further, the vessel is insulated, the system is not in a position to gain heat from or lose heat to its surroundings, i.e. no exchange of energy with its surroundings.

2. Closed System: A system which can exchange energy but not matter with its surroundings is called a closed system. In above equation, it is not isolated, then it is a closed system.

Another equation:

A chemical reaction taking place in a cylinder closed by a piston. Mixture of O_2 and H_2 in ratio of 1 : 2 by volume enclosed in a cylinder by a frictionless piston.

$$2H_2(g) + O_2\ (g) \rightarrow 2H_2\ O(l)\ \Delta H\ 285.9\ kJ$$

A large amount of heat is evolved.

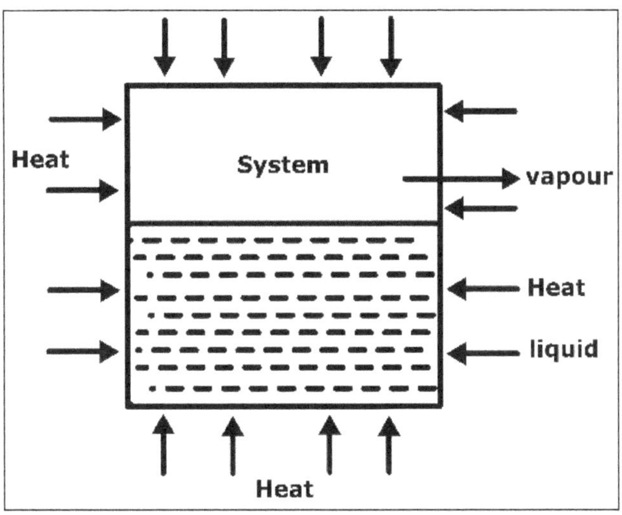

Closed system (A liquid in contact with its vapor in a vessel of constant Volume), not insulated.

During the reaction, the volume of the gases reaction mixture decreases as the reactions is completed by the fall in volume. So, the piston moves down due to the same work being done by the surroundings on the system. The energy is in the form of mechanical work has been transferred from the surroundings to the system. Thus, the system is in a position to exchange energy with its surroundings. But cannot exchange matter with surroundings as the cylinder remains closed throughout.

3. Open System: A system which can exchange matter as well as energy with its surroundings is said to be an open system. E.g. Expansion of water in an open beaker. The water in open beaker absorbs heat for evaporation from the surroundings and escapes into the surroundings as water vapour. The system exchange energy and matter with surface. Animals & plants are the examples of open system.

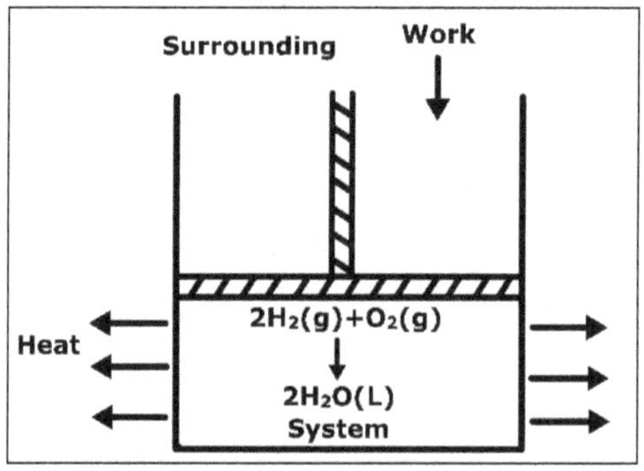

A reaction taking place in a cylinder enclosed by a system.

- Macroscopic Properties: The properties associated with a macroscopic system (consisting of large no. of particles) are called macroscopic properties. These properties are Pressure, Volume, Temperature, concentration, density, viscosity, surface tension, refractive index, color etc.

State of a System and State Variables

- State of a system: The state of a system is defined when its macroscopic properties are specified. When macroscopic properties of a system have definite values, the system is said to be in definite state. Whenever there is a change in any one of the macroscopic properties, the system is said to change into a different state.
- State variables: The macroscopic properties of a system are often called state variables, since the state of a system change with the change in any of the macroscopic properties. E.g. Pressure, Temperature, Volume, mass and composition

For defining a state of a system, it is not necessary to define the values of all state variables, since they are independent.

In homogenous system, these properties are inter-related by equation of state PV = RT, where P = pressure, V = Volume of 1 mole of gas, R = gas constant, T = absolute Temperature. The value of any of these P, V, T can be calculated provided those of other 2 are known. They are independent variables. Two variables generally defined are T (absolute Temperature) & Pressure they are independent variables. The third variable V is called dependent variable as its value depends on Temperature & Pressure. In closed system consisting of one or more components, mass is not a state variable.

State and Path Functions

Work and heat are path functions. Value of q & w depends upon how the process is carried out and which route is adopted,

$$\Delta U = \int_i^f dU$$

ΔU depends on initial and final states of system but is independent of path. Therefore, it is a state function (exact differential), But q cannot be written as $q_f - q_i$, it depends on path. Therefore, it is a path function (Inexact differential).

Zeroth Law of Thermodynamics

This law states that two systems in thermal equilibrium with a third system are also in thermal equilibrium with each other.

First Law of Thermodynamics

- Work: Work is defined as any quantity that flows across the boundary of a system during a change in its state and its affect is noted in the surrounding in the form of either weights are higher or lower in the surrounding.

Work done on the system = +ve

Work done by the system is = −ve.

Unit = J $\quad$ 1 J = 10^7 ergs

- Heat: Mode of energy exchange between the system and surroundings which results from the difference in the temperature of the system and the surroundings.

Unit – Calorie (A calorie is the quantity of heat required to raise the temperature one gram of H_2O through 1 °C).

If system has absorbed heat from its surroundings = +q.

If system has given heat to the surroundings = -q.

All other forms of energy tend to change ultimately into heat.

- Internal Energy: Every substance is associated with a definite amount of energy which depends upon its chemical nature as well as upon its T (absolute Temperature), Pressure & Volume. This energy is known as Internal energy. Exact magnitude of this energy is not known. Since, the chemical nature includes translational, rotational, vibrational movements of the molecule, nature of individual atoms, arrangement and number of electrons energy of nucleus etc.

Internal energy of a substance system is a definite quantity and is a function of only of the state (i.e. chemical nature, T (absolute Temperature), Pressure, Volume & composition) of the system at the given moment irrespective of the manner in which that state has been brought out.

Equivalence between Mechanical Work and Heat Work can be converted in to equivalent of heat energy.

W (Mechanical work) $\propto$ H(heat produced).

$\Rightarrow$ W = JH (J – Joule's mechanical equivalent of heat).

When H = 1 then W = J

J is defined as the amount of work done to produce the calorie of heat and is numerically = 4.184 or 4.184×10^7 ergs. (1 Calorie = 4.184 J).

First Law of Thermodynamics

Energy can neither be created nor destroyed, although it can transformed from one from to another.

Energy can be produced by the destruction of mass, the two quantities being related by the expression $E = mc^2$ (Einstein Equation).

Modified Law: The total mass and energy of an isolated system remains unchanged.

Mathematical Formulation of the First Law of Thermodynamics

Suppose a system is subjected to change of pressure and volume, let initial state be A and find state represented as B.

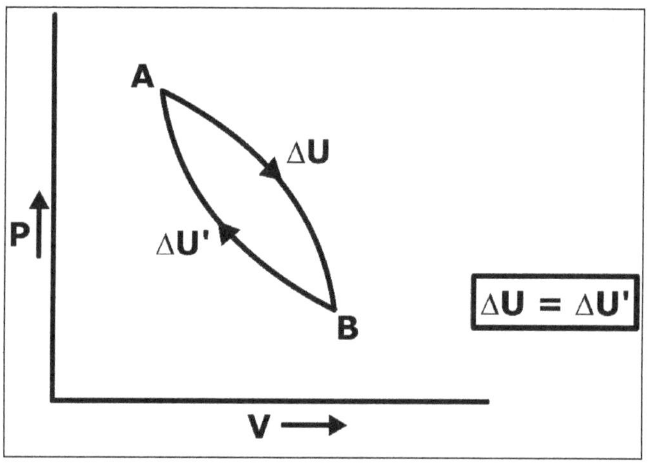

U_A & U_B - Energies associated with the system in its states A & B.

Change in internal energy: $\Delta U = U_B - U_A$

Energy change accompanying a process is a function only of the initial and final states of the system and is independent of the path by which the change is brought about.

Energy Changes in Relation to Work and Heat Changes

- $U_A \rightarrow$ Energy of system in its state A.

- $U_B \rightarrow$ Energy of system in its state B.

Suppose the system while undergoing change from states $A \rightarrow B$ absorbs heat 'q' from the surroundings and also performs some words (mechanical or electrical) equal to 'w'. The absorption of heat by the system tends to raise the energy of the system. The performance of work by the system tends to lower the energy of the system since it requires expenditure of energy.

$$\Delta U = U_B - U_A = q - w$$

In general, if in a given process, the quantity of heat transferred from surface to system is q and work done in the process is w.

$$\Delta U = q + w$$

- Work done by surroundings on the system $w \rightarrow$ +ve.
- Work done by system on the surroundings $w \rightarrow$ −ve.

State Functions

State function has a definite value for a particular state of the system. It is independent of the manner in which the state is reached. It depends only on initial and final states of the system and not on the path e.g. T (absolute Temperature), Pressure & Volume & dU.

$$w, q \rightarrow \text{Path function}$$

$$\int_{u_1}^{u_2} dU = U_2 - U_1$$

(exact diff.)

$$\text{But} \int_{q_1}^{q_2} dq \neq q_2 - q_1 \ \& \int_{w_1}^{w_2} dw \neq w_2 - w_1$$

(inexact diff.)

Enthalpy of a System

Let change of state of a system is brought about at constant P and there will be a change in Volume (V). Let increase in volume occurs from $V_A \rightarrow V_B$ at constant P.

Work done by the system, w:

$$w = -P(V_B - V_A)$$
$$\Delta U = q + w(\text{First law})$$

$$= q + (-P(V_B - V_A) = q - P(V_B - V_A))$$
$$\Rightarrow (U_B - U_A) = q - P(V_B - V_A)$$
$$\Rightarrow (U_B + PV_B) - (U_A - PV_A) = q$$

The quantity U + PV is called enthalpy and is designated by H.

H = U + PV

It represents the total energy stored in the system.

- Internal energy (U), Pressure, Volume are definite properties.
- H is also a definite property.

$$H_B - H_A = \Delta H = q$$

Since, $U_B - U_A = q - P(V_B - V_A)$ $(\because q = \Delta H)$

$\Delta U = q - P\Delta V$ or $\Delta H = \Delta U + P\Delta V$

Thermochemistry

The branch of chemistry which deals with the energy changes involved in chemical reaction is called thermochemistry. It is the study of energy and heat associated with chemical reactions. Thermochemistry focuses on the energy changes, primarily on the system's energy exchange with its surroundings. Thermochemistry can predict the quantities of reactants and products during the course of the reaction. It is also useful in prediction of the spontaneity of the reaction. It merges the concepts of thermodynamics with the concept of energy in the form of chemical bonds. The quantities like enthalpy, heat capacity, entropy, free energy, heat of formation are mainly calculated through this. According to thermochemistry, the change in energy which occurs in chemical reaction is mainly because of the change of bond energy, i.e., it results from the breaking of bonds in the reactants and formation of new bonds in products.

Thermochemistry is based on two laws:

- Lavoisier and Laplace's law: This law states that the change in energy accompanying any transformation is equal and opposite to change in energy accompanying the reverse process.
- Hess's law: This law states that the change in energy accompanying any transformation is same whether the process occurs in one step or many steps.

Lavoisier, Laplace and Hess have also done investigation on specific heat and latent heat.

Change in Thermodynamic Quantities during a Chemical Reaction

Change in Internal Energy of a Chemical Reaction

Consider a chemical reaction during which the temperature and volume is kept constant, i.e., dV=0. Thus, work done (w) is also equal to zero as w=PdV. Therefore, equation of the First law (viz., $\Delta U = q + w$) becomes:

$$\Delta U = q_v$$

where q_v stands for the heat exchanged at constant volume.

Let U_R be the internal energy of the reactants and U_P be the internal energy of the products, thus change in internal energy will be:

$$\Delta U = U_P - U_R = q_v$$

Change in Enthalpy of a Chemical Reaction

The heat exchanged at constant pressure is known as the enthalpy change. Suppose q_p be the heat exchanged during a chemical reaction which is occurring at constant pressure. Therefore,

$$\Delta H = q_p$$

Let H_R be the enthalpy of the reactants and H_P be the enthalpy of the products, thus change in enthalpy will be:

$$\Delta H = H_p - H_R = q_p$$

Thermochemistry enables us to predict the amount of heat that would be evolved or absorbed in a process without actually performing a tedious set of experiments in the laboratory. The energy changes for the processes which are not feasible experimentally can also be calculated through thermochemistry.

- Sign convention: Reactions in which heat is absorbed by the system are called endothermic reactions. In such reactions $H_P > H_R$, so ΔH is positive. Since the energy of the system also increases by the absorption of heat thus ΔU is also positive in endothermic reactions. While the reactions in which heat is evolved, are called exothermic reactions. In such reactions HP < HR so ΔH is negative. In such reactions ΔU is also negative.

Relation between Enthalpy at Constant Volume (q_v) and Enthalpy at Constant Pressure (q_p)

The relation between ΔH and ΔU is given by:

$$\Delta H = \Delta U + P\Delta V$$

where, ΔV is the volume change taking place in a reaction.

Since $q_v = \Delta U$ and $q_p = \Delta H$, therefore we can write equation $\Delta H = \Delta U + P\Delta V$ as $q_p = q_v + P\Delta V$

Now writing the above equation in simplified manner;

For n moles of an ideal gas,

$$PV = nRT$$

Suppose n1 be the number of moles for gaseous reactants and n_2 be the number of moles of gaseous products. Let n2 > n1. Thus increase in the number of moles is given by $n_2 - n_1 = \Delta n_g$. The corresponding increase in volume (ΔV) will be given by (V/n) Δn_g. Therefore,

$$P\Delta V = P(V/n)\Delta n_g = RT \times \Delta n_g$$

Thus, $P\Delta V = RT \times \Delta n_g$

Substituting equation $P\Delta V = P(V/n)\Delta n_g = RT \times \Delta n_g$ in equation $q_p = q_v + P\Delta V$,

$$q_p = q_v + \Delta n_g RT$$

In the above equation Δn_g stands for the difference between the number of moles of gaseous products and gaseous reactants.

Enthalpy of a Chemical Reaction

Standard Enthalpy change of a reaction is defined as the enthalpy change of reaction determined at 25 °C and at 1atm pressure and is denoted by ΔH°. Considering various enthalpy changes:

Enthalpy of Formation

The enthalpy of formation can be defined as the amount of heat exchanged at constant temperature and pressure during the formation of one mole of the substance from its constituent elements in their standard states. It is represented by $\Delta_f H$. Its unit is kJ mol^{-1}.

For example, the enthalpy of formation of CO_2 is equal to the enthalpy change for the following reaction:

$$C(s)+O_2(g)\rightarrow CO_2(g) \qquad \Delta_f H=-393.5 \text{ kJ mol}^{-1}$$

$$H_2(g)+\frac{1}{2}O_2(g)\rightarrow H_2O(l) \qquad \Delta_f H=-285.830 \text{ kJ mol}^{-1}$$

$$H_2O(l)\rightarrow H_2(g)+\frac{1}{2}O_2(g) \qquad \Delta_f H=+285.830 \text{ kJ mol}^{-1}$$

Enthalpy of Combustion

It is defined as the enthalpy change that takes place when one mole of a substance is burnt completely in the presence of oxygen at a given temperature and pressure. It is denoted by $\Delta_c H$ and the unit is kJ mol^{-1}. The combustion is always an exothermic process. For example, combustion of methane:

$$CH_4(g)+CO_2(g)\rightarrow CO_2(g)+2H_2O(l) \qquad \Delta_c H^\theta(298 \text{ K})=-890 \text{kJ mol}^{-1}$$

$$C_2H_2(g)+\frac{7}{2}O_2(s)\rightarrow 2CO_2(g)+3H_2O(l) \qquad \Delta H^\circ(298 \text{ K})=-1560 \text{ kJ mol}^{-1}$$

Enthalpy of Solution

The amount of heat exchanged when 1 mole of solute is dissolved in a sufficient amount of solvent at a specified temperature and pressure is known as enthalpy of solution.

For example,

$$HCl(g)+10\ H_2O(l)\rightarrow HCl.10H_2O(aq) \quad \Delta H=-69.01 \text{ kJ mol}^{-1}$$

$$HCl(g)+40\ H_2O(l)\rightarrow HCl.40H_2O(aq) \quad \Delta H=-72.79 \text{k J mol}^{-1}$$

These values of ΔH show the general dependence of the heat of solution on the amount of the solvent. As more and more solvent is used the value of heat of solution changes. As the amount of the solvent increases the resulting solution becomes more dilute and ultimately it becomes so dilute that further addition of solvent produces no enthalpy change. This solution is known as infinitely dilute solution.

Enthalpy of Sublimation

It is the amount of enthalpy change to convert one mole of a solid to vapor state at a given temperature and pressure.

$$H_2O(1)\rightarrow H_2O(g) \qquad \Delta_{sub}H^\theta(298 \text{ K})=50.0 \text{ kJmol}^{-1}$$

Enthalpy of Fusion

It is the change in enthalpy to convert one mole of a solid to its liquid state at a given temperature and pressure.

$$H_2O(s) \rightarrow H_2O(l) \qquad \Delta_{fus}H^\theta(298\ K) = 6.0\ kJmol^{-1}$$

Enthalpy of Atomization

It is the amount of heat required to convert one mole of a substance into its constituent atoms in the gaseous state.

$$C(graphite) \xrightarrow{atomization} C(g) \qquad \Delta_a H(C) = 716.68\ kJ\ mol^{-1}$$

$$H_2(g) \xrightarrow{atomization} 2H(g) \qquad \Delta_a H(H) = 436\ kJ\ mol^{-1}$$

Determination of Enthalpies of Reactions

Enthalpies of reactions at 25°C can be determined if ΔH°_f values of the reactants and products involved in the reactions are known as,

$$\Delta H^\circ = \Sigma \Delta H^\circ_f(\text{products}) - \Sigma \Delta H^\circ_f(\text{reactants})$$

By convention, ΔH°_f values for the elements in their standard states are taken as zero.

Kirchhoff Equation: Variation of Enthalpy of Reaction with Temperature

The change in enthalpy of any physical or chemical process varies with temperature at constant pressure. The effect of temperature on the enthalpy can be understood as follows:

Consider a reaction,

$$aA + bB \rightarrow cC + dD$$

The enthalpy change for the above reaction will be:

$$\Delta H = \Sigma H_{products} - \Sigma H_{reactants} = (cH_C + dH_D) - (aH_A + bH_B)$$

Differentiating equation above with respect to temperature, keeping pressure constant

$$\left[\frac{\partial(\Delta H)}{\partial T}\right]_P = \left(\frac{\partial H_C}{\partial T}\right)_P + d\left(\frac{\partial H_D}{\partial T}\right)_P - a\left(\frac{\partial H_A}{\partial T}\right)_P - b\left(\frac{\partial H_B}{\partial T}\right)_P$$

Since, $C_P = (\partial H / \partial T)_P$

Therefore, equation $\left[\frac{\partial(\Delta H)}{\partial T}\right]_P = \left(\frac{\partial H_C}{\partial T}\right)_P + d\left(\frac{\partial H_D}{\partial T}\right)_P - a\left(\frac{\partial H_A}{\partial T}\right)_P - b\left(\frac{\partial H_B}{\partial T}\right)_P$ can be written as,

$$\left[\frac{\partial(\Delta H)}{\partial T}\right]_P = \sum C_P(\text{products}) - \sum C_P(\text{reactants})$$

$$\left[\frac{\partial(\Delta H)}{\partial T}\right]_P = cC_{P,C} + dC_{P,D} - aC_{P,A} - bC_{P,B} = \Delta C_P$$

where ΔC_p = Sum of heat capacities of products – Sum of heat capacities of reactants. Equation above is known as Kirchhoff equation. This equation states that the variation of ΔH of a reaction with a temperature at constant pressure is equal to ΔC_p of the system, i.e.,

$$\left[\partial(\Delta H)/\partial T\right]_P = \Delta C_P$$

Rearranging the above equation,

$$d(\Delta H) = \Delta C_p dT$$

Similarly, the dependence of enthalpy on temperature at constant volume is given by,

$$\left[\partial(\Delta H)/\partial T\right]_V = \Delta C_V \text{ or } d(\Delta U) = \Delta C_V dT$$

If the temperature range is small, then change in heat capacity is given by, (assuming heat capacities are not dependent on temperature).

$$\int_{T_1}^{T_2} d(\Delta H) = \int_{T_1}^{T_2} \Delta C_p dT = \Delta C_p \int_{T_1}^{T_2} dT \text{ or } \Delta H_2 - \Delta H_1 = \Delta C_p (T_2 - T_1)$$

Similarly,

$$\Delta U_2 - \Delta U_1 = \Delta C_V (T_2 - T_1)$$

If the temperature range is not small then the heat capacities will vary with temperature. Thus it is convenient to express the heat capacity as a power series in Temperature (T) i.e.

$$C_P = \alpha + \beta T + \gamma T^2$$

where α β, and γ are constants for a given species. Similarly,

$$\Delta C_P = = \Delta\alpha + \Delta\beta T + \Delta\gamma T^2 + \ldots..$$

Substituting equation $\Delta C_P = = \Delta\alpha + \Delta\beta T + \Delta\gamma T^2 + ...$ in equation $\left[\partial(\Delta H)/\partial T\right]_P = \Delta C_P$ and integrating between T_1 and T_2, we get,

$$\int_{T_1}^{T_2} d(\Delta H) = \int_{T_1}^{T_2} \left(\Delta\alpha + \Delta\beta T + \Delta\gamma T^2\right) dT$$

Or,

$$\Delta H_2 - \Delta H_1 = \Delta\alpha(T_2 - T_1) + (1/2)\Delta\beta\left(T_2^2 - T_1^2\right) + \frac{1}{3}\Delta\gamma\left(T_2^3 - T_1^3\right)$$

Equation above is the integrated Kirchhoff equation.

Flame and Explosion Temperatures

The combustion of a gaseous fuel in air occurs so rapidly that the heat produced during combustion does not get enough time to dissipate into the surroundings. Thus, combustion process is found to be equivalent to an adiabatic process. The entire amount of heat produced is used up to heat the gases which are produced during combustion. Maximum flame temperature is defined as the maximum temperature attained by the flame zone (containing the resultant gases) due to the heat evolved by the combustion of the fuel under adiabatic conditions at constant pressure. On the other hand, if the combustion is carried out under adiabatic conditions at constant volume, the maximum temperature attained is called maximum explosion temperature.

Kirchhoff equation is used to calculate the maximum flame temperature for an isobaric adiabatic process. This is done as follows,

$$d(\Delta H)/dT = \Delta C_P \quad \text{or} \quad d(\Delta H) = \Delta C_P dT$$

Integrating the above equation gives,

$$\int d(\Delta H) = \Delta C_P \int_{T_i}^{T_f} dT \quad \text{or} \quad \Delta H = \Delta C_P\left(T_f - T_i\right)$$

In the above equation ΔCP is assumed to be constant that is why it is taken outside the integral sign. Thus, if the values of ΔH, ΔCP and the initial temperature T_i are known then the final temperature T_f (maximum flame temperature) can be calculated.

Hess's Law

Hess's law was established by the Russian chemist German H. Hess in 1840. This law is known as Hess's law of constant heat summation. This law states that the amount of heat evolved and absorbed in a process, including a chemical change, is the same whether the process takes place in one or several steps, i.e., total change in enthalpy does not change during the course of the reaction. Thus Hess's law is also known as principle of conservation of energy. The change in enthalpy does not depend on the path taken from the initial to the final state (i.e. enthalpy is a state function). Reaction enthalpy changes can be determined by calorimetry for many reactions. It is of particular utility in calculations of the heat of reactions which are difficult for practical calorimetric measurements. The overall energy needed for a chemical reaction can be determined by Hess's law.

Hess's law states that the enthalpy change (i.e., heat of reaction at constant pressure) in a chemical reaction does not depend on the path between the initial and final states of the system. That is the overall change in enthalpy is same during a chemical change of a reaction regardless of the number of steps through which the reaction has been taken place. For example, for a change from reactant to product that can take place in four steps or a single step, the total enthalpy change will be same.

Single step process:

Reactant → Product ΔH

Multiple step process:

Reactant → A ΔH_1

A → B ΔH_2

B → C ΔH_3

C → Product ΔH_4

According to Hess's law,

$$\Delta H = \Delta H_1 + \Delta H_2 + \Delta H_3 + \Delta H_4$$

This can also be shown by following diagram:

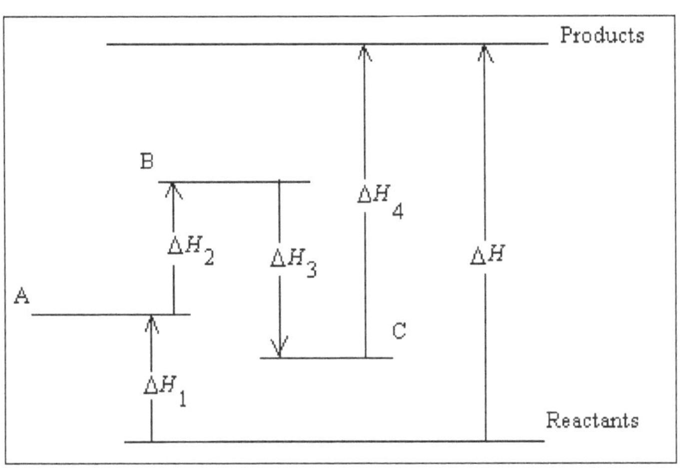

This generalization means that enthalpy of the reaction depends only on the initial reactants and the final products and not at all on the intermediate products that can be formed. Thus, enthalpy change which cannot be measured directly is calculated by Hess's law. If the net enthalpy change of the reaction is negative, then the reaction is said to be exothermic; positive value for enthalpy change corresponds to endothermic reactions.

Hess's law states that changes in enthalpy are additive. Thus for a single reaction change in enthalpy ΔH is given by:

$$\Delta H^\circ_{reaction} = \sum \Delta H^\circ_{f(products)} - \sum \Delta H^\circ_{f(reactants)}$$

where ΔH_f stands for the enthalpy of formation and superscripts ° represent standard state values. The above equation is the combination of two reactions. These are:

- Reactants → Elements

  $$\Delta H^\circ = -\sum \Delta^\circ_{f(reactants)}$$

- Elements → Products

  $$\Delta H^\circ = \sum \Delta^\circ_{f(products)}$$

Extension of Hess's Law

The changes in entropy and in Gibbs free energy can also be calculated by applying the concepts of Hess's law. The Bordwell thermodynamic cycle is an example of such an extension which takes advantage of easily measured equilibria and redox potentials to determine experimentally inaccessible Gibbs free energy values.

Thus the change in free energy can be determined by:

$$\Delta G^\circ_{reaction} = \sum \Delta G^\circ_{f(products)} - \sum \Delta G^\circ_{f(reactants)}$$

But entropy can be measured as an absolute value thus entropy of formation is not required, simply absolute values of entropy are used.

$$\Delta S^\circ_{reaction} = \sum \Delta S^\circ_{products} - \sum \Delta S^\circ_{reactants}$$

Applications of Hess's Law

Hess's law of constant heat summation is useful in the determination of enthalpies of the following:

- Calculation of enthalpies of reactions.

- Determination of enthalpy changes of slow reactions.
- Calculation of enthalpies of formation.
- Enthalpy of formation of reactive intermediates.
- It helps in determining the lattice energies of ionic substances by building Born-Haber cycles if the electron affinity to form the anion is known.

Example: The heat of dissociation per mole of a gaseous water at 18 °C and 1 atm is 241750 J, calculate its value at 68° C. Data given are:

$$C_P(H_2O) = 33.56;\ C_P(H_2) = 28.83;\ C_P(O_2) = 29.12\ JK^{-1}\ mol^{-1}$$

Solution: The dissociation reaction is,

$$H_2O(g) \rightarrow H_2(g) + \frac{1}{2}O_2(g) \qquad \Delta H°(291\ K) = 241750\ J$$

$$\Delta C_P = C_P(H_2) + \frac{1}{2}C_P(O_2) - C_P(H_2O)$$

$$= \left(28.83 + \frac{1}{2} * 29.12 - 33.56\right) J\ K^{-1}\ mol^{-1}$$

$$= 9.83\ J\ K^{-1}\ mol^{-1}$$

Therefore,

$$\Delta C_P * \Delta T = \left(9.83\, J K^{-1} mol^{-1}\right) * (50\ K)$$

$$= 491.5\ J\ mol^{-1}$$

$$\Delta H°(341\ K) = \Delta H°(291\ K) + C_P * \Delta T$$

$$= 241750 + 491.5 = 242241.5\ J\ mol^{-1}.$$

References

- Physical-chemistry, science: britannica.com, Retrieved 09, March 2020
- Electrochemistry: byjus.com, Retrieved 26, July 2020
- Chemical-kinetics, science: britannica.com, Retrieved 13, April 2020
- Chemical-equilibrium: byjus.com, Retrieved 07, February 2020

Permissions

All chapters in this book are published with permission under the Creative Commons Attribution Share Alike License or equivalent. Every chapter published in this book has been scrutinized by our experts. Their significance has been extensively debated. The topics covered herein carry significant information for a comprehensive understanding. They may even be implemented as practical applications or may be referred to as a beginning point for further studies.

We would like to thank the editorial team for lending their expertise to make the book truly unique. They have played a crucial role in the development of this book. Without their invaluable contributions this book wouldn't have been possible. They have made vital efforts to compile up to date information on the varied aspects of this subject to make this book a valuable addition to the collection of many professionals and students.

This book was conceptualized with the vision of imparting up-to-date and integrated information in this field. To ensure the same, a matchless editorial board was set up. Every individual on the board went through rigorous rounds of assessment to prove their worth. After which they invested a large part of their time researching and compiling the most relevant data for our readers.

The editorial board has been involved in producing this book since its inception. They have spent rigorous hours researching and exploring the diverse topics which have resulted in the successful publishing of this book. They have passed on their knowledge of decades through this book. To expedite this challenging task, the publisher supported the team at every step. A small team of assistant editors was also appointed to further simplify the editing procedure and attain best results for the readers.

Apart from the editorial board, the designing team has also invested a significant amount of their time in understanding the subject and creating the most relevant covers. They scrutinized every image to scout for the most suitable representation of the subject and create an appropriate cover for the book.

The publishing team has been an ardent support to the editorial, designing and production team. Their endless efforts to recruit the best for this project, has resulted in the accomplishment of this book. They are a veteran in the field of academics and their pool of knowledge is as vast as their experience in printing. Their expertise and guidance has proved useful at every step. Their uncompromising quality standards have made this book an exceptional effort. Their encouragement from time to time has been an inspiration for everyone.

The publisher and the editorial board hope that this book will prove to be a valuable piece of knowledge for students, practitioners and scholars across the globe.

Index

www.ingramcontent.com/pod-product-compliance
Lightning Source LLC
LaVergne TN
LVHW080612210926
848485LV00003B/145